Tales of the Tail

Tales of the Tail

Legendary Cats Throughout History

Solomon Raj

UNIEK ENTERPRISES

CONTENTS

INDEX

Introduction

INTRODUCTION

In the tremendous embroidered artwork of mankind's set of experiences, one animal has reliably transformed civilizations across the globe — the feline. Stories of the Tail set out on an excursion through time, investigating the incredible felines whose accounts have been woven into the actual texture of our social, otherworldly, and day to day existences. From the venerated gods of old Egypt to the devilish catlike joke artists in fables, and from fearless wartime allies to web sensations, this investigation dives into the rich and changed jobs that felines have played since the beginning of time.

1.1 The Getting through Appeal of Felines

Felines have enamored the human creative mind for centuries, their puzzling presence frequently moving wonder, respect, and, on occasion, a dash of strange notion. The intrinsic tastefulness, autonomy, and strange nature of these animals have made them loved mates as well as images of divine nature, sorcery, and versatility.

1.2 The Worldwide Meaning of Felines

While the homegrown feline, Felis catus, has turned into a natural presence in families around the world, its importance goes past the limits of homes and urban communities. From the deserts of Egypt to the clamoring roads of Tokyo, felines have wound up at the focal point of social practices, strict convictions, and imaginative articulations. Stories of the Tail try to unwind the general allure of felines and their different jobs in forming the human experience.

Felines in Antiquated Civilizations

2.1 Egyptian Felines: Gatekeepers and Divinities

In antiquated Egypt, felines were something beyond pets — they were venerated and revered. We investigate the hallowed job of felines in Egyptian culture, where they were related with the goddess Bastet, the catlike divinity of home, ripeness, and security. Felines became images of beauty and guardianship, with numerous families keeping them as appreciated partners.

2.2 Felines in Mesopotamia: Confounding Defenders

In the support of progress, Mesopotamian societies additionally respected felines. We dig into the social and otherworldly meaning of felines in Mesopotamia, where they were accepted to have supernatural characteristics and were related with defensive gods. The adoration for felines in this antiquated locale molded their representative significance in craftsmanship, writing, and day to day existence.

Cat Comedians in Legends

3.1 European Folktales: Puss in Boots and Then some

European legends is rich with stories of crafty and wicked felines. We disentangle the narratives of Puss in Boots and other cat pranksters who outmaneuver their direction to fortune and acclaim. These stories investigate the original of the sharp feline and its getting through claim in molding social accounts.

3.2 Asian Old stories: The Nekomata and Bakeneko

In Asian societies, especially in Japan, old stories acquaints us with the enchanted Nekomata and Bakeneko — felines with heavenly capacities. We dig into the legends encompassing these shape-moving cats, investigating their associations with the soul world and their job in people convictions.

Felines in Wartime

4.1 Old Felines in Fight

All through old developments, felines wound up on the forefronts of fighting. From filling in as nuisance regulators in antiquated Egypt to giving solace to warriors in old Rome, we investigate the jobs of felines in the midst of contention. Their presence in disaster areas features the exceptional and changed commitments of cat associates over the entire course of time.

4.2 Wartime Felines: Sidekicks and Legends

In later times, felines play played huge parts in wartime endeavors. We reveal the accounts of wartime felines who filled in as allies to warriors, offered close to home help, and, surprisingly, identified approaching assaults. These catlike legends feature the profound bonds framed among people and their bold feline colleagues on the combat zone.

Felines and Otherworldly Images

5.1 Hinduism: Felines as Friends of Goddesses

In Hinduism, felines are related with different goddesses and are viewed as favorable partners. We investigate the emblematic meaning of felines in Hindu folklore, their part in strict ceremonies, and the heavenly associations ascribed to these elegant animals.

5.2 Celtic Practices: Pixie Felines and Watchmen

In Celtic old stories, felines are accepted to have otherworldly characteristics, filling in as watchmen and allies to the pixie society. We dive into the stories of pixie felines and their jobs in Celtic customs, where these magical cats explore the domains between the seen and concealed.

Felines in Workmanship and Writing

6.1 Old and Renaissance Craftsmanship: Felines as Images

The imaginative portrayal of felines has advanced throughout the long term, from antiquated Egyptian symbolic representations to Renaissance works of art. We investigate how felines became images in workmanship, addressing subjects of secret, style, and the heavenly. Prestigious craftsmen have deified felines in their works, abandoning a tradition of cat propelled imagination.

6.2 Scholarly Felines: From Familiars to Heroes

In writing, felines play taken on different parts, from filling in as familiars in supernatural practices to becoming heroes by their own doing. We disentangle the scholarly tradition of felines, investigating their depiction in works going from Edgar Allan Poe's strange stories to Lewis Carroll's unusual undertakings.

Felines in Present day Culture

7.1 Web Felines: From Viral Sensations to Social Symbols

The computerized age has introduced another time for felines, changing them into web sensations and social symbols. We investigate the peculiarity of web felines, from Testy Feline's notable glower to Nyan Feline's rainbow-mixed flight. These catlike famous people have caught the hearts of millions, forming the contemporary view of felines in mainstream society.

7.2 Felines in Worldwide Craftsmanship and Design

In the advanced imaginative scene, felines keep on rousing imagination and development. Craftsmen and planners attract upon cat class to make a combination of conventional imagery and present day style. We analyze how felines are addressed in worldwide craftsmanship and style, their impact rising above social limits.

Felines in Various Societies and Their Emblematic Implications

8.1 Asian Societies: Maneki-neko and Then some

In Asian societies, felines hold assorted representative implications. We investigate the notable Maneki-neko, the enticing feline, and its job as an image of favorable luck. Also, we dive into the meaning of felines in Chinese and Korean customs, revealing the layers of importance related with these perplexing animals.

8.2 Caribbean Legends: Jumbee Felines and Otherworldly Protection

In Caribbean legends, felines assume the job of otherworldly safeguards and defenders. We unwind the narratives of Jumbee Felines and their importance in warding off heavenly substances. The otherworldly characteristics credited to these felines mirror the social extravagance of Caribbean convictions.

Felines in Notions and Black magic

9.1 Notions Encompassing Felines

Since forever ago, felines have been both loved and dreaded, prompting a heap of strange notions. We investigate the notions encompassing felines, from convictions about their otherworldly powers to the scandalous relationship with misfortune. These stories mirror the complex and in some cases problematic mentalities toward felines in different societies.

9.2 Felines in Black magic: Familiars and Enchanted Partners

In the domain of black magic and the mysterious, felines have frequently been portrayed as familiars and enchanted partners to witches. We dive into the authentic associations among felines and black magic, investigating the imagery, fables, and social insights that have formed the supernatural connection among witches and their catlike mates.

Genuine Legends and Famous Felines

10.1 Genuine Legends: Felines in Search and Salvage

A few felines rise above the domain of legend to turn out to be genuine legends. We investigate the tales of felines associated with search and salvage activities, exhibiting their remarkable capacities, preparing, and commitments to saving lives. These catlike legends exhibit the exceptional effect felines can have in the midst of emergency.

10.2 Famous Felines: Stories of Feline Thieves and Devilish Cats

While certain felines are praised for their bravery, others are notorious for their naughty endeavors. We unwind the stories of feline robbers and famous cat miscreants, investigating the narratives of shrewd felines who transformed history through robbery, naughtiness, and at times recovery.

The Concealed Universe of Felines

11.1 Felines and the Paranormal

Felines have for quite some time been related with the paranormal, with stories of their capacity to detect spirits and powerful energies. We investigate the persona encompassing felines and the concealed world, from their presence in antiquated legends to contemporary records of felines displaying odd way of behaving.

11.2 The Mystic Association: Felines and Human Feelings

The natural idea of felines has prompted convictions in a mystic association among cats and people. We dive into accounts of felines showing a significant comprehension of human feelings, offering solace, friendship, and, surprisingly, going about as daily reassurance creatures in the midst of hardship.

The Eventual fate of Cat Legends

12.1 Felines in the Mechanical Age

As we step into a period overwhelmed by innovation, the job of felines keeps on advancing. We investigate how felines are adjusting to the advanced age, from their presence via virtual entertainment stages to developments in pet consideration innovation. The crossing point of customary imagery and current headways shapes the continuous account of felines in the mechanical age.

12.2 Preservation and Feline Government assistance

While felines play played different parts over the entire course of time, their prosperity is presently a developing concern. We examine the significance of protection endeavors, mindful pet possession, and drives zeroed in on guaranteeing the government assistance of felines around the world. The developing connection among people and felines requires a decent way to deal with safeguard both the social importance and the moral treatment of these wonderful animals.

1. **A Brief Overview of the Cultural Significance of Cats**

Felines, with their cryptic presence and effortless disposition, play rose above the part of simple homegrown allies to become social symbols with significant representative implications across human advancements. This investigation dives into the multi-layered social meaning of felines, unwinding their jobs in religion, old stories, craftsmanship, writing, and day to day existence. From the worshipped cat divinities of old Egypt to the notions encompassing dark felines and the contemporary impact of web feline images, this outline gives an exhaustive look into the rich embroidery of human-cat connections.

1.1 The Feline as an Image of Secret and Freedom

At the core of the social meaning of felines lies their natural secret and freedom. Felines, in contrast to many trained creatures, hold a feeling of independence and independence that has caught the human creative mind for a really long time. This inborn capacity to move smoothly through both homegrown and wild scenes adds to their emblematic portrayal as animals of secret, opportunity, and slippery excellence.

1.2 Felines as Friends and Healers

Past their secretive charm, felines have been loved as friends, giving comfort, solace, and friendship to people. The quieting murmur of a satisfied feline has been connected to pressure decrease and mending properties, encouraging a remarkable connection among cats and their human partners. This double job of secret and friendship frames the underpinning of the social meaning of felines.

Antiquated Egypt - Felines as Heavenly Creatures

2.1 Bastet, the Goddess of Home and Security

In antiquated Egypt, felines were not just pets; they were adored as heavenly creatures. Bastet, the feline goddess, was revered for her relationship with home, richness, assurance, and smooth savagery. We investigate the social meaning of felines in old Egyptian culture, their part in strict practices, and their raised status as images of both home life and heavenly power.

2.2 Felines in Egyptian Craftsmanship and Pictographs

The social meaning of felines in antiquated Egypt is clearly reflected in craftsmanship and pictographs. Felines were highlighted noticeably in paintings, models, and even burial places, exhibiting their significance in both day to day routine and life following death. The careful portrayals of felines in Egyptian craftsmanship give experiences into the adored status of these creatures and their basic job in the social texture of old Egypt.

European Legends - Felines as Puzzling Joke artists

3.1 Puss in Boots and Cat Comedians

European legends is rich with stories of cat joke artists, like the scandalous Puss in Boots. We dig into the old stories encompassing felines as smart and crafty sidekicks who utilize their mind to explore through life's difficulties. These accounts investigate the duality of felines as both naughty joke artists and faithful partners.

3.2 Dark Felines and Odd notions

In European old stories, dark felines have been dependent upon both positive and negative notions. While certain societies view them as images of best of luck, others partner them with black magic and odd notions. We disentangle the social subtleties encompassing dark felines, investigating their representative importance in different European practices.

Felines in Asian Societies - Images of Fortune and Astuteness

4.1 Maneki-neko: The Coaxing Feline

In Asian societies, especially in Japan, the Maneki-neko, or enticing feline, is a notable image of favorable luck and success. We investigate the starting points of this notorious puppet and its far and wide presence in organizations and families. The Maneki-neko addresses the positive social meaning of felines, carrying karma and accomplishment to the people who embrace its presence.

4.2 Felines in Chinese and Korean Customs

Felines assume assorted parts in Chinese and Korean societies, representing security, karma, and even everlasting status. From antiquated legends to contemporary convictions, we analyze how felines are woven into the texture of these societies, becoming both legendary animals and cherished associates.

Felines in Craftsmanship and Writing - Images of Tastefulness and Secret

5.1 Antiquated and Renaissance Workmanship: Felines as Images

The creative portrayal of felines has developed over hundreds of years, from the emblematic portrayals in antiquated Egyptian craftsmanship to the Renaissance works of art that caught the polish and secret of cats. We investigate how felines became getting through images in visual expressions, addressing effortlessness, sexiness, and the confounding idea of these animals.

5.2 Scholarly Felines: From Familiars to Heroes

In writing, felines have taken on different jobs, from enchanted familiars to heroes by their own doing. We disentangle the abstract tradition of felines, analyzing their depiction in works going from Shakespearean plays to current books. Felines in writing frequently typify both the homegrown and the strange, mirroring the social interest with these multi-layered animals.

Felines in Wartime - Sidekicks and Mascots

6.1 Old Felines in Fight

Since the beginning of time, felines have went with people into the turbulent conditions of fighting. From old civic establishments to later contentions, felines have filled in as colleagues, giving solace to troopers and procuring their place as esteemed mascots. We investigate the jobs of felines in wartime, revealing insight into the one of a kind bonds framed between these creatures and the people who served close by them.

6.2 Wartime Felines: Legends and Survivors

In the midst of war, felines have shown astounding versatility and boldness. We uncover accounts of wartime felines who became legends, whether through making fighters aware of approaching assaults or offering close to home help on the forefronts. The social meaning of wartime felines reaches out past their commonsense jobs, representing trust and friendship despite misfortune.

Felines in Present day Culture - Web Sensations and Worldwide Symbols

7.1 Web Felines: From Viral Sensations to Social Symbols

The approach of the web has launch felines into worldwide fame, with viral sensations like Grouchy Feline and Nyan Feline becoming social symbols. We investigate the peculiarity of web felines, their effect on mainstream society, and how they have changed the view of felines in the computerized age. The social meaning of present day felines reaches out past customary mediums, impacting patterns, images, and, surprisingly, forming social communications.

7.2 Felines in Worldwide Workmanship and Style

Felines keep on moving imagination in the contemporary craftsmanship and style scene. From very good quality design to road craftsmanship, we dig into the portrayals of felines in worldwide imaginative articulations. The persevering through appeal of felines as images of class and secret tracks down new translations in the steadily advancing domains of workmanship and style.

Odd notions and Black magic - Felines Among Fantasy and Reality

8.1 Notions Encompassing Felines

The notions encompassing felines are pretty much as different as the way of life that hold them. From convictions in the mysterious powers of felines to the notorious relationship with misfortune, we investigate the social notions that have formed view of felines since forever ago. The polarity of felines as both captivating and premonition animals is reflected in these well established convictions.

8.2 Felines in Black magic: Familiars and Enchanted Partners

In the domain of black magic and the mysterious, felines have frequently been portrayed as familiars and enchanted partners to witches. We dive into the authentic associations among felines and black magic, investigating the imagery, old stories, and social insights that have formed the otherworldly connection among witches and their catlike mates. The social meaning of felines in black magic mirrors a complicated transaction of dread, interest, and worship.

Felines in Regular daily existence - Pets and Social Symbols

9.1 Felines as Well known Pets

In the contemporary world, felines have gotten their place as adored pets in huge number of families. We investigate the ordinary social meaning of felines as buddies, consistent encouragement creatures, and individuals from the family.

The connection among people and their catlike mates adds to the persevering through ubiquity of felines as pets.

9.2 The Feline Bistro Peculiarity

The rise of feline bistros addresses an exceptional convergence of culture, business, and friendship. We dive into the feline bistro peculiarity, looking at how these foundations have become social centers where individuals can connect with felines in a loose and remedial climate. The social meaning of feline bistros reaches out past the pleasure in espresso and baked goods, cultivating a more profound appreciation for the friendship of cats.

2. The Bond Between Humans and Cats Throughout History

The connection among people and felines is an embroidery woven with strings of friendship, secret, and shared understanding that stretches across centuries. From the hallowed sanctuaries of old Egypt to the comfortable corners of present day front rooms, the connection among people and felines has risen above social, geological, and transient limits. This investigation sets out on a thorough excursion through history, following the development of this one of a kind relationship, unwinding the social, emblematic, and reasonable perspectives that have formed the getting through connection among people and felines.

1.1 The Starting points of the Human-Cat Bond

The early sections of human development set up for the rise of the human-cat bond. As people progressed from traveling ways of life to settled horticultural networks, the presence of felines became interlaced with the rural scene. We investigate the archeological proof and early social portrayals that enlighten the underlying strings of association among people and felines.

1.2 The Persona of Felines in Antiquated Societies

In old developments like Egypt, Mesopotamia, and Greece, felines held a venerated status that went past reasonable jobs. The supernatural characteristics credited to felines in strict convictions, fables, and craftsmanship formed the impression of these catlike mates as puzzling creatures. We dig into the hallowed imagery and social meaning of felines in antiquated societies, looking at how they became the two partners and images of heavenly security.

Felines in Old Egypt - Gatekeepers of The hereafter

2.1 Bastet, the Catlike Divinity

In old Egypt, felines rose to divine status through their relationship with the goddess Bastet. We investigate the strict meaning of felines in Egyptian culture, where they were loved as defenders of homes and watchmen of life following death.

The many-sided connection among Egyptians and their catlike associates reveals insight into the significant social and otherworldly jobs felines played in this antiquated human progress.

2.2 Felines as Images in Egyptian Craftsmanship and Writing

The social meaning of felines in old Egypt is strikingly portrayed in workmanship and writing. Felines were highlighted noticeably in symbolic representations, figures, and even as allies to pharaohs. We unwind the imagery of felines in Egyptian imaginative articulations, digging into how their polish and secret were deified in the visual and composed accounts of the time.

Felines in Center Eastern Societies - From Mesopotamia to Islam

3.1 Felines in Mesopotamian Convictions

In the rich bow of Mesopotamia, felines assumed basic parts in strict convictions and regular daily existence. We investigate how Mesopotamian societies saw felines as defenders against malicious spirits and bugs, stressing their viable and emblematic importance. The social transaction among people and felines in this area established the groundwork for the getting through bond that would reverberate through the ages.

3.2 Felines in Islamic Practices

As Islam spread across the Center East, felines kept on being esteemed and regarded creatures. The lessons of Prophet Muhammad featured the graciousness and thought owed to felines, encouraging a humane disposition toward these catlike partners. We research how Islamic customs formed the treatment of felines, making a social heritage that rose above strict limits.

Felines in Old Greece and Rome - Images of Family life and Security

4.1 Old Greece: Felines in Folklore and Day to day existence

In old Greece, felines were coordinated into both folklore and day to day existence. From the catlike mates of the goddess Artemis to their pragmatic jobs in controlling vermin, we investigate the multi-layered connection among Greeks and felines. The social subtleties encompassing felines in antiquated Greece uncover their double jobs as images of family life and defenders of homes.

4.2 Rome: Felines as Gatekeepers of Grain

The Romans acquired a profound appreciation for felines from their Greek ancestors. We look at how felines in old Rome were regarded for their part in safeguarding grain stores from rodents. The commonsense worth of felines, joined with their representative significance, highlighted the extraordinary association among Romans and their catlike partners.

Felines in Archaic Europe - Strains and Changes

5.1 Felines and the Archaic Church

The archaic period in Europe saw complex cooperations between felines, odd notions, and the Congregation. Felines, once connected with agnostic convictions, became entrapped in the doubts and fears of the Congregation. We investigate the strains encompassing felines during the Medieval times, inspecting how they explored changing social scenes.

5.2 Felines as Partners and Vermin Regulators

In spite of the difficulties looked during bygone eras, felines continued as significant friends and gifted bother regulators. The advantageous connection among people and felines thrived in the homegrown settings of palaces, religious communities, and families. We reveal the jobs of felines in regular daily existence during middle age Europe, featuring their versatility and flexibility.

The Renaissance and Felines - Images of Style and Family life

6.1 Felines in Renaissance Workmanship

The Renaissance period saw a restoration of interest in old style workmanship and a reconnection with nature. Felines, with their effortlessness and tastefulness, became well known subjects in Renaissance craftsmanship. We investigate how specialists portrayed felines in artworks and models, hoisting them to images of refinement and homegrown magnificence.

6.2 Felines in Homegrown Settings

In Renaissance families, felines kept on assuming imperative parts as friends and defenders of homes. We dig into the developing social mentalities toward felines during this period, featuring their presence in homegrown settings and the rise of a more agreeable connection among people and their catlike companions.

Felines in Asia - Gatekeepers, Associates, and Social Images

7.1 Felines in Asian Craftsmanship and Writing

In different Asian societies, felines have been commended in workmanship and writing for their tastefulness and imagery. From Chinese canvases to Japanese haiku, we investigate how felines became social images, exemplifying characteristics of excellence, secret, and friendship. The creative portrayals of felines in Asia mirror a profound appreciation for their stylish and social importance.

7.2 Felines in Japanese Fables

Japanese fables acquaints us with mysterious felines with heavenly capacities. The Nekomata and Bakeneko are legendary animals that typify the social subtleties of felines in Japan. We disentangle the stories of these shape-moving cats, analyzing how they became both loved and dreaded in Japanese customs.

Felines in the Time of Investigation - Pilgrims and Marine Associates

8.1 Felines on Boats

As travelers set out on oceanic undertakings during the Time of Investigation, felines became basic friends on board delivers. We investigate the down to earth jobs of boat felines in controlling vermin and lifting team confidence level. The social trade among mariners and their catlike shipmates made a permanent imprint on the nautical history of felines.

8.2 Felines as Worldwide Explorers

The worldwide goes of pilgrims and dealers worked with the spread of felines across mainlands. Felines adjusted to different conditions, becoming essential pieces of networks around the world. We research what the development of felines during this period meant for their social importance and set their status as worldwide partners.

Felines in the Edification - Illumination Thoughts and Really impacting Points of view

9.1 Felines in Edification Workmanship and Writing

The Edification period achieved a change in scholarly and social points of view. Felines kept on being portrayed in craftsmanship and writing, reflecting changing mentalities toward creatures and homegrown life. We investigate the portrayals of felines during the Edification, underlining their parts in molding the feel of the time.

9.2 Felines in Privileged Salons

Felines tracked down their direction into the salons of the Edification, becoming allies to scholarly people and blue-bloods. We dive into the social elements of felines in these scholarly center points, where they gave both solace and motivation to the illuminators of the time. The presence of felines in refined salons features their persevering through bid among people of impact.

Felines in the Modern Unrest - Urbanization and Changing Elements

10.1 Felines in Metropolitan Conditions

The Modern Unrest denoted a time of quick urbanization, and felines adjusted to the changing scenes of urban areas. We investigate how felines became associates in progressively jam-packed metropolitan conditions, exploring the difficulties and open doors introduced by the modern shift. The advancing jobs of felines reflected the changes happening in the public arena.

10.2 Felines in Writing and Mainstream society

Writing and mainstream society of the Modern Unrest kept on highlighting felines as adored mates. From Dickensian stories to early artistic depictions, we inspect how felines stayed necessary to social accounts during this extraordinary period. The narratives of felines in writing mirror the persevering through associations among people and their catlike companions.

Felines in the twentieth 100 years - Wars, Social Moves, and Friend Creatures

11.1 Felines in Wartime

The twentieth century achieved worldwide contentions that profoundly affected human social orders. Felines, as in past conflicts, assumed parts as partners, mascots, and even legends. We investigate the wartime encounters of felines, both on the home front and the war zones, featuring their flexibility and the comfort they gave during seasons of conflict.

11.2 Felines in the Ascent of The suburbs

The post-war time saw the ascent of rural residing, and felines kept on being appreciated individuals from families. We analyze what the social movements of the mid-twentieth century meant for the elements of the human-cat bond in rural settings. Felines became basic to the pure picture of rural life, giving friendship to families in the changing scenes of the period.

Felines in the Advanced Age - From Web Sensations to Worldwide Symbols

12.1 Felines on the Web

The approach of the web launch felines into worldwide fame. We investigate the peculiarity of web felines, from viral sensations to social symbols. Images, recordings, and online entertainment stages turned into the new materials for depicting the appeal and shenanigans of felines. The computerized age changed the manner in which we see felines as well as hardened their place as persevering through images of online culture.

12.2 Felines in Contemporary Living

In the 21st hundred years, felines have flawlessly coordinated into the texture of contemporary living. We analyze the assorted jobs of felines in present day families, where they keep on being esteemed partners, basic reassurance creatures, and even forces to be reckoned with via virtual entertainment. The social meaning of felines has developed with the times, adjusting to the speedy, interconnected scenes of contemporary living.

Chapter 1

Ancient Purr-suasions

In the records of old history, in the midst of the great embroidered artwork of human civilization, one animal arose as both sidekick and god, leaving a permanent paw print on the pages of time — the feline. From the spiritualist terrains of old Egypt to the baffling domains of Mesopotamia, felines in times long past held an importance that rose above simple taming. This investigation, crossing 3000 words, digs into the enamoring universe of "Old Murmur suasions," unwinding the complex jobs of felines in old social orders, where they were respected as images of heavenliness, watchmen of eternity, and baffling allies to mankind.

1. **Divine Buddies: Felines in Antiquated Egypt**
   1.1 Bastet, the Catlike God
   Our process starts in the midst of the enormous pyramids and the rich banks of the Nile, where the old Egyptians raised the feline to a heavenly status. At the core of this catlike worship was Bastet, the goddess of home, ripeness, and insurance. Frequently portrayed with the top of a lioness or a homegrown feline, Bastet epitomized the characteristics that Egyptians respected in their catlike partners — elegance, dexterity, and a wild defensive impulse.
   The old Egyptians accepted that felines were the natural signs of Bastet, and consequently, they held a sacrosanct spot in families. Hurting a feline, even incidentally, was met with extreme outcomes, mirroring the profound respect Egyptians had for these baffling animals. In this segment, we'll investigate the strict customs, creative portrayals, and regular routines of antiquated Egyptians entwined with the heavenly quintessence of their catlike colleagues.
   1.2 Felines in Ceremonies and Entombments
   Past the walls of homes, felines assumed a fundamental part in strict functions. Sanctuaries committed to Bastet were embellished with feline sculptures and pictures, mirroring the goddess' considerate and defensive nature. Aficionados offered preserved felines as recognitions, accepting that this act would conjure

the goddess' approval and protect their homes from hurt.

Felines were additionally conspicuous in the funerary acts of antiquated Egypt. The Book of the Dead, a manual for existence in the wake of death, highlighted spells and outlines portraying felines as gatekeepers of the departed.

These catlike defenders were accepted to go with the spirits through the dangerous excursion of life following death, representing the persevering through association among people and felines past the domain of the living.

2. Gatekeepers of the Hidden world: Felines in Mesopotamian Folklore

2.1 The Mesopotamian Association

As we navigate the old scenes, we show up in Mesopotamia, where the way of life of Sumerians, Babylonians, and Assyrians thrived. Around here, felines expected a particular job in the complicated embroidery of folklore and day to day existence. Dissimilar to the Egyptians, the Mesopotamians didn't revere felines, yet their imagery and presence were similarly significant.

2.2 Marduk and the Consecrated Feline

In Mesopotamian folklore, felines were frequently connected with Marduk, the preeminent divine force of the Babylonian pantheon. Marduk's sacrosanct creature, the mušḫuššu, was a legendary animal portrayed as a mythical beast or snake with the top of a cat. This baffling combination represented both savagery and security, typifying the duality intrinsic in felines.

The mušḫuššu was noticeably highlighted in the unbelievable Ishtar Entryway of Babylon, a heavenly construction embellished with dynamic blue tiles portraying the hallowed animal. This door filled in as both a physical and emblematic section highlight the city, underscoring the defensive job of the mušḫuššu and, likewise, felines in the more extensive Mesopotamian perspective.

3. Felines in Greek and Roman Domains: Murmur sonifications of Beauty and Secret

3.1 Artemis and the Feline's Beauty

The social impact of felines stretched out to old Greece and Rome, where they became related with gods exemplifying effortlessness, hunting ability, and secret. In Greek folklore, Artemis, the goddess of the chase, wild, and wild creatures, was frequently joined by felines. These animals, with their covertness and spryness, reflected the characteristics Artemis esteemed in her heavenly interests.

Felines, especially wildcats, became images of untamed nature and the puzzling powers that existed outside human ability to comprehend. The Greeks respected the catlike capacity to explore both the wild and the homegrown circle, perceiving a duality suggestive of the subtle idea of old divine beings and goddesses.

3.2 Roman Veneration for Felines

In antiquated Rome, felines were regarded for their rat getting capacities, adding to their presence in both metropolitan and provincial settings.

The Roman goddess of opportunity, Libertas, was frequently portrayed with a feline next to her, representing the freedom to uninhibitedly wander and explore spaces.

Felines were additionally connected with the goddess Diana, the Roman partner to the Greek Artemis. As a gatekeeper of the wild, Diana's proclivity for felines stressed their job as allies to the people who track the borderlands among progress and the untamed.

4. Felines in Asian Societies: Watchmen, Signs, and Fortune

4.1 Felines in Chinese Legends

Our investigation currently moves toward the East, where felines play held different parts in the social accounts of Asian civilizations. In Chinese fables, felines have been both adored and dreaded. The incredible "Feline God" or "Li Shou" is a legendary figure accepted to control rodents and mice, filling in as a gatekeeper of storage facilities and homes. This altruistic translation of felines lines up with their pragmatic use as nuisance regulators in rural social orders.

On the other hand, Chinese notions partner dark felines with misfortune and disaster. This double discernment mirrors the complex transaction between social convictions and the functional utility of felines in various settings.

4.2 Maneki-neko: The Alluring Feline of Japan

In Japan, the Maneki-neko, or alluring feline, has turned into a notorious image of favorable luck and thriving. Portrayed by a raised paw, the Maneki-neko is accepted to draw in sure energy and riches. These puppets, frequently found in organizations and homes, grandstand the social combination of reasonableness and imagery that felines encapsulate in Japanese culture.

The Maneki-neko's enticing motion and kindhearted articulation have changed felines into specialists of flourishing, rising above their job as simple allies to become images of karma and overflow.

5. Felines in Contemporary Setting: From Mews to Images

5.1 Felines in Present day Old stories and Odd notions

The impact of old murmur suasions stretches out into contemporary times, where felines keep on being the subject of old stories and odd notions. While the relationship of felines with witches has blurred, a few notions persevere. Dark felines, regardless of their unfortunate underlying meanings in Western societies, are viewed as images of best of luck in different areas of the planet. This determination of convictions features the persevering and developing nature of human-feline connections.

5.2 Web Felines and Images

In the advanced age, felines have flawlessly changed from antiquated images to web sensations. The ascent of feline images, filled by the openness of online stages, has transformed normal felines into worldwide big names. Images like Irritable Feline, Nyan Feline, and the incalculable viral feline recordings circling the web have become present day articulations of the persevering through interest with these catlike sidekicks.

The web's capacity to interface individuals internationally has changed felines from confined images to widespread symbols, rising above social limits and making a common appreciation for the jokes and appeal of our catlike companions.

1.1 Cats in Ancient Egyptian Culture

In the domain of old Egypt, where the Nile Stream streamed with nurturing waters and transcending pyramids repeated the glory of a high level progress, one animal ruled — the feline. In this investigation crossing 2000 words, we dig into the hallowed meaning of felines in old Egyptian culture, disentangling their heavenly jobs, strict imagery, and the significant effect these cryptic animals had on the regular routines and otherworldly convictions of individuals of the Nile.

1. **Bastet, the Catlike Divinity**

   1.1 The Goddess of Home and Assurance

   At the core of the Egyptian respect for felines stood Bastet, the goddess who exemplified the characteristics Egyptians appreciated in their catlike sidekicks. Bastet, frequently portrayed with the top of a lioness or a homegrown feline, held influence over home, ripeness, and security. Her kindness stretched out to the homegrown circle, and families accepted that having a feline in their home would conjure the goddess' approval.

   In the hallowed city of Bubastis, a focal point of love devoted to Bastet, stupendous sanctuaries rose to pay tribute to the catlike god. Pioneers and aficionados rushed to these sanctuaries, bringing contributions and looking for the insurance and gifts that Bastet was accepted to present to her adherents.

   1.2 Felines as the Natural Signs of Bastet

   The antiquated Egyptians went past simple adoration for Bastet; they accepted that homegrown felines were the natural signs of the goddess herself. Felines, with their beauty, deftness, and sharp hunting abilities, typified the characteristics related with Bastet. Subsequently, hurting a feline, even unintentionally, was viewed as a grave offense, deserving of regulation. This love for felines was not bound to the home; it reached out into the cultural and legitimate domains, underscoring the significant association between the heavenly and the cat in antiquated Egyptian culture.

2. **Felines in Customs and Entombments**

   2.1 Sanctuaries Decorated with Cat Sculptures

   Sanctuaries devoted to Bastet were embellished with sculptures and portrayals of felines. These catlike portrayals filled in as both strict images and gatekeepers of the consecrated spaces. The presence of these sculptures not just supported the association among felines and the heavenly yet additionally made a climate of respect and security inside the sanctuary walls.

2.2 Preserved Contributions to Bastet

Aficionados communicated their dedication to Bastet through intricate ceremonies, frequently including embalmed felines presented as recognitions. These embalmed felines were accepted to convey the requests and wishes of individuals to the goddess, filling in as mediators between the human domain and the heavenly. The act of offering embalmed felines likewise mirrored the significance of the catlike god in the strict and otherworldly existences of the antiquated Egyptians.

2.3 Felines in Funerary Practices

The social meaning of felines stretched out into the domain of life following death. In the Book of the Dead, a manual for the excursion through life following death, outlines frequently portrayed felines as defenders of the departed. The conviction that felines went with spirits through the hazardous excursion of eternity built up their job as gatekeepers and guides, accentuating the persevering through connection among people and felines that rose above the limits of life and demise.

3. ## Felines as Defenders and Emblematic Watchmen

3.1 Watchmen Against Abhorrent Spirits

Past their strict jobs, felines were venerated as defenders against pernicious powers. Egyptians accepted that the presence of a feline in the home would avert underhanded spirits and carry endowments to the family. Felines, with their sharp faculties and nighttime exercises, were viewed as careful sentinels making preparations for concealed dangers that prowled in the shadows.

3.2 Feline Ornaments and Images of Assurance

The defensive characteristics ascribed to felines tracked down articulation in the production of feline special necklaces. These little charms, frequently made of valuable metals or stones, were worn as defensive charms. The special necklaces included many-sided carvings or etchings of felines, representing the heavenly security and positive energy related with these catlike mates.

4. ## The Feline in Day to day existence

4.1 Felines as Friends and Nuisance Regulators

While the strict and representative parts of felines were vital to antiquated Egyptian culture, felines were additionally appreciated as regular friends. In a transcendently horticultural society, felines assumed a functional part as compelling bug regulators, especially in shielding grain stores from rodents and mice.

The training of felines filled a double need, meeting the down to earth necessities of horticulture while giving close to home and otherworldly satisfaction to individuals. Felines became essential individuals from Egyptian families, acquiring their keep through their normal hunting impulses while charming themselves to their human friends.

4.2 Felines in Workmanship and Writing

The significant impact of felines penetrated the imaginative articulations of old Egypt. Felines were highlighted in paintings, figures, and adornments, deified in different structures that caught their effortlessness and persona. These creative portrayals served as recognitions for the catlike sidekicks as well as for of protecting their social importance for a long time into the future.

In writing, as well, felines tracked down a spot in the social story. Sonnets and stories frequently praised the ideals of felines, depicting them as faithful, canny, and adored individuals from the family. The stories of felines in day to day existence added a nuanced layer to the social texture, displaying the different jobs these animals played in the hearts and homes of the old Egyptians.

5. Heritage and Congruity: Felines in Current Egypt

5.1 Current Veneration for Felines

The veneration for felines in Egypt didn't blur with the ways of the world; it persevered, developing into a cutting edge appreciation for these animals. While the strict and otherworldly viewpoints might have changed, the social meaning of felines stays tangible in contemporary Egyptian culture.

In the clamoring roads of Cairo and the calmer corners of towns, felines keep on wandering unreservedly, embraced by a culture that perceives their worth as the two colleagues and watchmen. The presence of homeless felines is met with resistance and, much of the time, care from the local area — a demonstration of the persevering through tradition of the old Egyptian love for these perplexing animals.

5.2 Felines in Workmanship, Writing, and Mainstream society

The imaginative tradition of felines perseveres in current Egyptian culture. Contemporary craftsmen draw motivation from antiquated portrayals, making lively and reminiscent works that give proper respect to the catlike associates of old.

Writing, as well, keeps on investigating the ageless connection among people and felines, catching the embodiment of the old social account in present day narrating.

In mainstream society, felines remain images of beauty, secret, and friendship. The web time has brought the shenanigans and appeal of Egyptian felines to a worldwide crowd, with virtual entertainment stages displaying the getting through allure of these animals in the computerized age.

1.2 Bastet, the Egyptian Cat Goddess

In the pantheon of antiquated Egyptian divinities, none spellbound the hearts and brains of individuals very like Bastet, the feline goddess. Worshipped for her elegance, defensive nature, and relationship with home and richness, Bastet arose as an image of heavenly cat energy, making a permanent imprint on the social, strict, and creative scene of old Egypt. In this investigation of 1900 words, we dive into the multi-layered parts of Bastet, unwinding the folklore, strict importance, and persevering through tradition of this cryptic Egyptian goddess.

1. **Bastet in Folklore: The Development of a Catlike Divinity**
   1.1 Beginnings and Affiliations
   Bastet's starting points can be followed back to the early dynastic time of old Egypt, where she at first showed up as a lioness-headed goddess. Her name, "Bastet," is accepted to get from the old Egyptian word "bast," signifying "devourer." As a lioness, Bastet epitomized savage insurance and was related with the forceful parts of a lion's temperament.

   Over the long run, in any case, the symbolism and affiliations encompassing Bastet went through a change. As the Egyptian development advanced, so did the impression of the goddess. The savagery of the lioness gave way to a more kindhearted picture, with Bastet expecting the type of a lioness or, all the more regularly, a homegrown feline. This development reflected the change in cultural perspectives, stressing the supporting and defensive characteristics of the homegrown feline, which charmed itself to the Egyptian public.

   1.2 Bastet and Sekhmet: Double Parts of the Catlike Godlikeness
   Bastet's advancement is personally attached to another catlike god, Sekhmet. Initially thought to be two particular goddesses, Bastet and Sekhmet later became perceived as double parts of a similar cat divine nature. While Bastet addressed the delicate, sustaining parts of the feline, Sekhmet exemplified its savage and horrendous characteristics.

   This duality displayed the multi-layered nature of felines in Egyptian culture. Bastet, as a homegrown feline, represented insurance, richness, and home, while Sekhmet, as a lioness, typified the champion and forceful viewpoints related with the chase.

   Together, these double perspectives portrayed the catlike heavenliness' effect on both the homegrown and wild domains.

2. **Bastet as the Defender of Home and Family**
   2.1 The Watchman of Homegrown Spaces
   One of Bastet's essential jobs was that of a watchman god, explicitly safeguarding the home and family. Families across old Egypt respected her as a generous power, guaranteeing the security and prosperity of families. Pictures and sculptures of Bastet were ordinarily positioned in homes as defensive charms, representing the heavenly presence that looked after the occupants.

   The homegrown feline's way of behaving, with its sharp faculties and careful nature, was firmly lined up with the defensive characteristics credited to Bastet. This association between the catlike friend and the goddess made a substantial connection between the heavenly and the daily existences of the Egyptian public.

   2.2 Bastet and Parenthood
   Bastet's defensive nature reached out to her job as a lady of richness and parenthood. The goddess was much of the time portrayed supporting cats or nursing

them, representing her association with the pattern of life, birth, and sustaining. Ladies looking for fruitfulness and safe labor went to Bastet for direction and endowments, stressing the goddess' impact over the patterns of life inside the Egyptian family structure.

In this maternal viewpoint, Bastet's tenderness and supporting characteristics stood out from the furious maternal senses of Sekhmet, offering a reasonable portrayal of the double idea of cat energy in both the homegrown and wild circles.

3. **Bastet in Strict Practices and Love**

3.1 Sanctuaries Devoted to Bastet

The love of Bastet was not restricted to the confidential circle; it stretched out to stupendous sanctuaries committed to the goddess. The most prominent of these sanctuaries was situated in the city of Bubastis, where Bastet was the important divinity. The city's noticeable quality as a focal point of love for the feline goddess hardened her significance in the strict and social scene of old Egypt.

Lovers ran to the sanctuaries of Bastet, bringing contributions and looking for the goddess' approval. Journeys to Bubastis during yearly celebrations became lively festivals, highlighting music, dance, and eats to pay tribute to the catlike divinity. The celebrations filled in as shared articulations of appreciation and dedication, building up the necessary job Bastet played in the strict acts of the Egyptian public.

3.2 Contributions and Ceremonies

Contributions to Bastet included preserved felines, which were introduced as badge of dedication and as solicitations for the goddess' security and gifts. The confidence in the otherworldly meaning of felines, as natural signs of Bastet, penetrated these customs. The preserved felines, painstakingly ready and embellished, filled in as courses between the human domain and the heavenly, representing the persevering through connection between individuals and their catlike defenders.

Ceremonies additionally included petitions and songs committed to Bastet, praising her excellencies and looking for her kindheartedness. The musical serenades and melodic psalms reverberated through the sanctuary corridors, supporting the profound association between the admirers and the catlike goddess.

4. **Bastet in Craftsmanship and Imagery**

4.1 Iconography of Bastet

The creative portrayal of Bastet reflected the goddess' double nature as both a homegrown and wild cat god. In her homegrown viewpoint, Bastet was portrayed with the top of a lioness or, all the more generally, a homegrown feline. Her stance frequently conveyed a feeling of effortlessness and supporting, underlining the tenderness and defensive characteristics related with the goddess.

Pictures of Bastet frequently included her holding an ankh, the image of life, featuring her association with ripeness and the patterns of life. Different portrayals exhibited the goddess with a sistrum, an instrument related with satisfaction and festivity, underscoring the happy parts of her love.

4.2 Bastet in Adornments and Ornaments

The imagery of Bastet reached out past sanctuaries and figures into daily existence. Egyptians embellished themselves with gems and special necklaces highlighting the picture of Bastet, conveying her defensive energy with them all through their everyday exercises. Feline formed special necklaces, cut from different materials like valuable stones or metals, filled in as charms that represented both otherworldly association and individual security.

The broad utilization of Bastet's symbolism in private embellishments featured the goddess' impact on individual lives, underlining the longing for her endowments and the apparent defensive characteristics innate in the catlike heavenly nature.

5. Tradition of Bastet: Cat Energy Through the Ages

5.1 Downfall and Resurgence

As the fortunes of old Egypt came and went, so did the conspicuousness of Bastet in the social and strict scene. With the ascent of various lines and moving strict convictions, the love of Bastet experienced times of decline. Nonetheless, the tradition of the feline goddess persevered, woven into the texture of Egyptian cognizance.

In later times, there has been a resurgence of interest and love for Bastet. The interest with old Egyptian culture, combined with a developing appreciation for the emblematic characteristics of felines, has prompted a restored investigation of the goddess' effect on contemporary profound practices.

5.2 Bastet's Impact Past Egypt

The tradition of Bastet reaches out past the boundaries of old Egypt, pervading worldwide awareness. The appeal of the feline goddess has risen above time and social limits, catching the minds of the people who track down reverberation in the imagery of cat elegance, assurance, and sustaining energy.

In mainstream society, Bastet's impact should be visible in writing, craftsmanship, and, surprisingly, current otherworldly practices. Creators and craftsmen draw motivation from the old portrayals of the goddess, integrating her imagery into contemporary accounts that investigate subjects of security, womanliness, and the secrets of life.

1.3 Cats as Symbols of Protection and Good Fortune

In the rich embroidery of mankind's set of experiences, scarcely any animals have involved a more venerated and puzzling situation than felines. Past their job as adored friends, felines have arisen as strong images of security and favorable luck in societies crossing the globe. This investigation, including 1700 words, digs into the social meaning of felines as images, unwinding the strings of legends, strange notion, and profound convictions that have raised these catlike creatures to the situation with gatekeepers and carriers of thriving.

1. **Old Egypt: Bastet's Heavenly Assurance**
   1.1 Bastet, the Egyptian Feline Goddess
   Our process starts in the old sands of Egypt, where the catlike persona arrived at its pinnacle. At the core of Egyptian veneration for felines stood Bastet, the goddess of home, fruitfulness, and insurance. Bastet, frequently portrayed with the top of a lioness or a homegrown feline, typified the supporting and defensive characteristics related with these confounding animals.

   The antiquated Egyptians accepted that homegrown felines were the natural indications of Bastet herself. Hurting a feline, even coincidentally, was viewed as a grave offense, as it was viewed as an attack against the heavenly goddess. Bastet's kindheartedness stretched out past the sanctuary walls and into the homes of individuals, making an unmistakable connection between the catlike gatekeepers and the day to day routines of the Egyptians.

   1.2 Felines in Egyptian Homes and Burial chambers
   The training of felines in antiquated Egypt was not simply functional; it was mixed with otherworldly importance. Families kept felines in their homes, for their bug control capacities as well as defenders against pernicious powers. The watchful and nighttime nature of felines made them regular gatekeepers, and their presence in families was accepted to shield the home from detestable spirits. In the domain of life following death, felines highlighted conspicuously in funerary practices. The Book of the Dead, a manual for the excursion through life following death, portrayed felines as partners and defenders of the departed. The faith in felines as guides in the hereafter built up their job as profound gatekeepers, expanding their defensive impact past the human domain.

2. **China: The Alluring Feline and Success**
   2.1 Maneki-neko: The Coaxing Feline of Japan
   Our investigation presently moves to the Far East, where the Maneki-neko, or coaxing feline, has turned into a famous image of favorable luck and thriving in Japanese culture. Portrayed by a raised paw and an enticing motion, the Maneki-neko is a typical sight in organizations and homes across Japan.

   The different varieties and embellishments enhancing the Maneki-neko convey explicit implications. A feline with a raised right paw is said to draw in riches and achievement, while a raised left paw is accepted to draw in clients and clients. The meaning of the varieties, for example, white for virtue and gold for abundance, adds layers of imagery to this enchanting cat puppet.

   2.2 Chinese Old stories: The Feline God and Assurance from Detestable Spirits
   In Chinese old stories, felines have been related with assurance from detestable

spirits and hardship. The "Feline God" or "Li Shou" is a legendary figure accepted to control rodents and mice, filling in as a gatekeeper of storage facilities and homes. The encouraging implications of the Feline God line up with the useful utility of felines in controlling nuisances, changing them into images of security and flourishing in Chinese culture.

In homes and organizations, Chinese old stories proposes that setting a feline doll close to the entry can avoid underhanded spirits and draw in amazing good fortune. This confidence in the defensive and promising characteristics of felines has persevered through ages, mirroring the getting through imagery of cat guardianship.

3. **Norse Folklore: Freyja's Chariot and Emblematic Effortlessness**
3.1 Freyja, the Norse Goddess of Adoration and Excellence

In the domains of Norse folklore, felines held an extraordinary spot as partners of Freyja, the goddess of adoration, excellence, and ripeness. Freyja was frequently portrayed with a chariot that was supposed to be pulled by enormous felines, underscoring their relationship with elegance and freedom.

Felines in Norse folklore represented both the homegrown and nature. Their capacity to explore between the two domains mirrored the duality of Freyja herself — a goddess who epitomized both the tenderness of affection and the strength of a hero. The presence of felines in Freyja's escort added a layer of imagery to their picture as effortless and puzzling creatures.

3.2 Felines as Emblematic Mates and Defenders

The association among Freyja and her catlike colleagues went past simple imagery. Norse individuals accepted that having a feline in the home could bring the gifts and favor of Freyja. Felines, with their full concentrations eyes and night-time propensities, were viewed as defenders against malicious powers, adjusting them to the job of watchmen in Norse families.

4. **Japanese Old stories: Nekomata and Extraordinary Watchmen**
4.1 Nekomata: Otherworldly Felines in Japanese Old stories

In Japanese old stories, felines are images of favorable luck as well as figures of otherworldly power. The Nekomata, a legendary feline with a forked tail, is accepted to have otherworldly capacities and extraordinary characteristics. These felines are said to can talk, shape-shift, and even control the spirits of the dead. While the Maneki-neko addresses the generous and promising characteristics of felines, the Nekomata exemplifies the strange and magical angles. This duality mirrors the complex idea of felines in Japanese fables, where they can be the two defenders and puzzling creatures with extraordinary powers.

4.2 Bakeneko and the Extraordinary Idea of Felines

Bakeneko, one more powerful feline in Japanese old stories, is accepted to be a homegrown feline that changes into an otherworldly animal with mysterious capacities. These felines are related with shapeshifting and frequently depicted with otherworldly powers, for example, clairvoyance and the capacity to control objects.

The extraordinary idea of Bakeneko adds a component of secret to the imagery of felines in Japanese old stories. While the Maneki-neko welcomes favorable luck, and the Nekomata has otherworldly abilities, the Bakeneko epitomizes the possibility that felines can exist on the limit between the common and the remarkable.

5. **Present day Setting: Felines in Contemporary Imagery**

5.1 Web Felines and the Computerized Age

As we change to the advanced time, the imagery of felines has taken on new aspects in the age of the web. The ascent of feline images and viral feline recordings has changed these catlike animals into worldwide images of humor, euphoria, and diversion. Images like Crotchety Feline and Nyan Feline have become social peculiarities, displaying the getting through interest with felines in the computerized age.

The web has given felines another stage to catch the aggregate creative mind, rising above social limits and making a common worldwide appreciation for the appeal and charm of these fuzzy sidekicks. In the computerized domain, felines keep on representing security and favorable luck as well as the cheerful and clever parts of life.

5.2 Treatment Felines and Consistent reassurance

In contemporary society, felines play likewise tracked down a part as treatment creatures, offering profound help and friendship to people confronting different difficulties. Treatment felines, similar to their canine partners, offer solace and comfort in medical care settings, showing the helpful advantages of the catlike human bond.

The mitigating presence of felines and their capacity to lessen pressure and nervousness have added to their representative job as everyday reassurance creatures. This cutting edge understanding adds another layer to the longstanding imagery of felines as defenders and bearers of positive energy.

Chapter 2

Cats in Mythology And Folklore

In the records of folklore and legends, the catlike presence arises as a charming and confounding power. Felines, with their smooth developments, penetrating eyes, and strange disposition, have woven themselves into the texture of social accounts across the globe. This investigation crossing 1900 words digs into the legendary and folkloric portrayals of felines, unwinding the strings of imagery, wizardry, and adoration that have raised these animals to the situation with unbelievable gatekeepers.

1. **Old Egypt: Bastet's Heavenly Buddies**
   1.1 Bastet, the Feline Goddess
   Our process begins in the sun-kissed terrains of old Egypt, where the catlike persona arrived at unmatched levels. At the core of Egyptian veneration for felines stood Bastet, the goddess of home, richness, and insurance. Portrayed with the top of a lioness or a homegrown feline, Bastet epitomized the sustaining and defensive characteristics related with these confounding animals.

   The old Egyptians accepted that homegrown felines were natural indications of Bastet herself. Hurting a feline, even incidentally, was viewed as a grave offense, as it was seen as an attack against the heavenly goddess. Bastet's kindheartedness reached out past the sanctuary walls and into the homes of individuals, making an unmistakable connection between the catlike gatekeepers and the regular routines of the Egyptians.

   1.2 Felines in Fanciful Jobs
   Past their relationship with Bastet, felines assumed significant parts in Egyptian folklore. The goddess Sekhmet, frequently portrayed as a lioness, addressed the fierce and damaging part of cat energy. Felines, in both homegrown and wild structures, became emblematic portrayals of the duality intrinsic in the heavenly, mirroring the complex idea of these animals in Egyptian folklore.

   Felines were additionally accepted to go with spirits on their excursion to eternity. In the Book of the Dead, outlines frequently portrayed felines as defenders and guides in the risky excursion through the hidden world. This fanciful job highlighted the profound meaning of felines, rising above their natural presence and interweaving with the texture of existence in the wake of death.

2. **Norse Folklore: Freyja's Catlike Associates**

2.1 Freyja, the Norse Goddess of Adoration and War

Our investigation presently navigates the domains of Norse folklore, where felines partook in a unique spot in the escort of Freyja, the goddess of adoration, magnificence, and richness. Freyja, frequently portrayed with a chariot drawn by huge felines, exemplified the combination of elegance and strength credited to cat creatures.

Felines in Norse folklore represented both home life and ferocity, typifying the double idea of Freyja herself — a goddess who could be both delicate and wild. The consideration of felines in Freyja's fanciful story added a component of secret and beauty to their emblematic presence in Norse culture.

2.2 Felines as Emblematic Mates

The relationship among Freyja and her catlike mates went past simple imagery. Norse individuals accepted that having a feline in the home could summon the gifts and favor of Freyja. Felines, with their full concentrations eyes and night-time propensities, were viewed as defenders against malicious powers, adjusting them to the job of watchmen in Norse families.

Felines, as emblematic mates of Freyja, became interlaced with thoughts of adoration, excellence, and the secrets of the heavenly. Their presence added a layer of charm to the fanciful scene of the Norse pantheon.

3. **Japanese Old stories: Nekomata and Otherworldly Creatures**

3.1 Nekomata: Otherworldly Felines in Japanese Old stories

In the embroidery of Japanese old stories, felines take on both considerate and puzzling personas. The Nekomata, a legendary feline with a forked tail, is accepted to have enchanted capacities and otherworldly characteristics. These felines are said to have the ability to talk, shape-shift, and even control the spirits of the dead.

The Nekomata's persona adds a component of the heavenly to the emblematic portrayal of felines in Japanese fables. While certain felines, similar to the Maneki-neko, represent favorable luck and success, the Nekomata exemplifies the strange and otherworldly parts of cat energy.

3.2 Bakeneko: Shapeshifting Felines

Bakeneko, one more otherworldly feline in Japanese legends, is accepted to be a homegrown feline that changes into a heavenly animal with mystical capacities. These felines are related with shapeshifting and frequently depicted with extraordinary powers, for example, clairvoyance and the capacity to control objects.

The extraordinary idea of Bakeneko adds a component of wizardry to the imagery of felines in Japanese old stories.

These shapeshifting cats obscure the limits between the common and the unprecedented, exemplifying the conviction that felines can exist in domains past the unremarkable.

4. **Celtic Folklore: Feline Sídhe and Supernatural Watchmen**

4.1 Feline Sídhe: Supernatural Pixie Felines

In the cloudy domains of Celtic folklore, the Feline Sídhe arises as a puzzling and powerful figure. Frequently portrayed as an enormous, dark feline with a white spot on its chest, the Feline Sídhe is accepted to be a pixie animal with otherworldly capacities. In Celtic legends, these felines were remembered to have the ability to take the spirits of the dead as they passed into the great beyond.

The Feline Sídhe's relationship with the pixie domain added an ethereal and enchanted aspect to the imagery of felines in Celtic folklore. Their presence was both dreaded and respected, highlighting the conviction that felines could act as courses between the human and soul universes.

4.2 Gatekeeper Spirits and Representative Edges

Celtic folklore additionally depicted felines as gatekeepers and defenders. Felines, with their sharp faculties and capacity to move quietly, were viewed as gatekeepers of the home and representative edges between various domains. Their careful presence added a layer of mysterious security to Celtic families, epitomizing the possibility that felines could see and explore supernatural energies.

5. **Center Eastern Old stories: Felines and Jinn**

5.1 Felines as Defenders Against Jinn

In Center Eastern old stories, felines are frequently connected with security against malignant heavenly substances known as Jinn. The conviction that felines can distinguish the presence of Jinn and avert their impact is well established in the social texture of many Center Eastern social orders.

Felines, with their sharp faculties and nighttime propensities, are accepted to be cautious gatekeepers against the concealed powers of the Jinn. Their job as defenders adds a functional and profound aspect to the emblematic portrayal of felines in Center Eastern fables.

5.2 Felines in Islamic Practice

Felines likewise track down a spot in Islamic practice, where they are viewed as perfect creatures. The Prophet Muhammad is said to have had an incredible affection for felines, and stories flourish of his communications with these animals. The regard for felines in Islamic culture is reflected in the lessons that stress thoughtfulness and care for creatures, including felines.

The positive relationship of felines with security against heavenly powers and their neatness lines up with the more extensive imagery of felines as gatekeepers in Center Eastern old stories.

6. **Current Legends: Dark Felines and Odd notions**

6.1 Dark Felines and Odd notions

In current legends, dark felines have become images of odd notions, both positive and negative. While certain societies accept that dark felines bring best of luck, others partner them with awful signs and black magic. The fables encompassing dark felines is frequently entwined with ideas of secret, wizardry, and the otherworldly.

The conviction that encountering a dark feline gets either fortune or disaster endures different societies. Strange notions encompassing dark felines feature the getting through persona and emblematic weight ascribed to these catlike creatures in contemporary fables.

6.2 Felines in Wiccan and Agnostic Practices

Inside present day agnostic and Wiccan practices, felines are frequently connected with supernatural energies and profound understanding. The paradigm of the "natural," a mysterious friend that helps professionals in their art, is habitually typified by felines. The faith in felines as enchanted and natural creatures lines up with their authentic jobs as gatekeepers and images of powerful associations.

Felines, especially dark felines, are some of the time viewed as soul guides and partners in enchanted rehearses. Their presence in present day agnostic and Wiccan customs mirrors a coherence of the representative significance of felines in enchanted and profound practices.

2.1 Cats in Norse Mythology

In the tremendous embroidered artwork of Norse folklore, where divine beings, monsters, and other fantastical creatures track the domains of presence, felines arise as captivating and mysterious figures. Past their everyday presence in day to day existence, felines in Norse folklore assume parts going from colleagues of divinities to gatekeepers of the great beyond. This investigation, traversing 2000 words, digs into the diverse jobs of felines in the Norse universe, disentangling the strings of imagery, wizardry, and holy affiliations that tight spot these catlike creatures to the mythic account.

1. Freyja's Chariot: Felines as Mates of the Goddess of Affection

1.1 Freyja: Love, Richness, and War

Our excursion into the Norse mythic domain starts with Freyja, quite possibly of the most noticeable goddess in the Norse pantheon. Frequently hailed as the goddess of adoration, richness, and war, Freyja possesses a focal situation in Norse folklore. Her relationship with magnificence, sexiness, and wizardry puts her in a class of heavenly figures that epitomize both elegance and power.

1.2 Felines in Freyja's Chariot

An eminent part of Freyja's folklore is her chariot, which is supposed to be drawn by two enormous felines. This unmistakable symbolism highlights the special association among Freyja and cat creatures. The imagery of felines pulling Freyja's chariot adds a component of persona to her persona, featuring the juxtaposition of the delicate and the wild parts of cat nature.

The consideration of felines in Freyja's escort fills in as a demonstration of the venerated status of these animals in Norse folklore. Felines, known for their freedom and beauty, line up with the characteristics related with Freyja herself. The mythic chariot drawn by these catlike partners emphasizes the entwined predeterminations of goddess and feline in the Norse cosmology.

2. **The Norse Homegrown Feline: Watchman of the Home**

2.1 Felines as Watchmen and Mates

Notwithstanding their relationship with the heavenly, homegrown felines held importance in the daily existences of Norse individuals. Felines were not simply pets; they were viewed as important mates and defenders of families. Their full concentrations eyes, sharp detects, and capacity to move quietly made them appropriate to the job of gatekeepers.

The Norse public, similar as different societies, appreciated the functional commitments of felines in controlling vermin. Nonetheless, the Norse perspective blessed felines with a more profound importance. Their cautious and nighttime nature adjusted them to characteristics of discernment, assurance, and secret, making them regarded individuals from Norse families.

2.2 Felines in Norse People Convictions

Norse legends further enhanced the imagery of felines as defenders. It was accepted that a feline present in a family could act as a watchman against noxious spirits and heavenly substances. The careful look of the feline was remembered to avoid fiendish powers, adding a layer of otherworldly importance to their part in Norse homes.

The Norse public's regard for the mysterious characteristics of felines stretched out past the reasonable items of irritation control. Felines, in their homegrown structure, became entwined with ideas of guardianship, making a social worship that recognized the catlike's one of a kind spot in the Norse family.

3. **The Catlike Imagery in Norse Enchantment and Black magic**

3.1 Felines in Seiðr: The Norse Mysterious Practice

Enchantment held a noticeable spot in Norse culture, and the act of seiðr, a type of Norse wizardry, involved a scope of customs and methods. Felines, with their apparent enchanted characteristics, became related with seiðr, further hardening their mysterious status in Norse society.

Seiðr experts, known as seiðkonur (sorceresses) or seiðmenn (magicians), were accepted to saddle mysterious powers for different purposes. Felines, frequently connected with shape-moving and extraordinary capacities, were believed to be familiars or soul guides for those rehearsing seiðr. The association among felines and wizardry added a layer of extraordinary imagery to the catlike creatures in the Norse otherworldly practice.

3.2 Felines as Shape-Shifters and Mysterious Partners

Felines as shape-shifters highlighted unmistakably in Norse folklore and legends.

Change was a repetitive subject in the mythic stories, and felines were accepted to have the capacity to expect various structures. The shape-moving nature of felines adjusted them to the enchanted and liquid nature of the mystical expressions in Norse culture.

The relationship of felines with shape-moving additionally stretched out to the conviction that witches and magicians could change into felines. This double personality supported the view of felines as liminal creatures, ready to navigate the limits between the common and the uncommon, the unremarkable and the mystical.

4. **Felines and The great beyond: Gatekeepers of the Dead**

4.1 Felines in Norse Funerary Practices

Norse convictions about the great beyond integrated the presence of felines as watchmen and colleagues of the departed. In funerary practices, felines were now and again covered close by people as faithful aides in the excursion to the domain of the dead. The imagery of felines as friends in the hereafter stressed their persevering through job past the limits of mortal presence.

The Book of the Dead, a manual for the great beyond, portrayed felines as defenders and guides in the excursion through the hidden world. Their supernatural relationship with the domains past highlighted the confidence in the otherworldly meaning of felines in Norse funerary ceremonies.

4.2 Society Convictions and Strange notions About Felines and Demise

Norse legends flourished with odd notions and convictions about the association among felines and passing. It was believed that a feline sitting on a bed during a disease could either retain the disorder or envoy the methodology of death. The feline's apparent aversion to the heavenly added an air of secret and premonition to its presence in snapshots of disease and demise.

Felines were additionally accepted to have the capacity to see spirits and extraordinary elements. Their way of behaving, particularly during the evening, was remembered to mirror their association with concealed powers. This otherworldly responsiveness added to the relationship of felines with the mysterious elements of the Norse universe.

5. **Current Viewpoints: Felines in Contemporary Norse Culture**

5.1 Felines in Norse-Enlivened Craftsmanship and Writing

The getting through imagery of felines in Norse folklore keeps on reverberating in contemporary Norse-propelled craftsmanship and writing. Craftsmen and creators draw upon the rich mythic legacy to depict felines as magical creatures, shape-shifters, and buddies of strong divinities. The notorious symbolism of Freyja's chariot attracted by felines frequently tracks down articulation current translations of Norse topics.

Scholarly works propelled by Norse folklore, like Neil Gaiman's "American Divine beings" or Joanne Harris' "The Good news of Loki," investigate the persevering through appeal of Norse divinities and legendary creatures, infrequently including felines as indispensable characters. These cutting edge portrayals add to the continuous interest with Norse folklore and its mysterious occupants.

5.2 Felines in Mainstream society and Norse Revivalism

In mainstream society, the emblematic reverberation of felines in Norse folklore has saturated different types of media. From online gatherings committed to Norse revivalism to the reception of Norse-propelled images in contemporary black magic and agnosticism, the mysterious characteristics of felines persevere as a wellspring of motivation.

Felines, frequently celebrated for their autonomy and strange appeal, keep on being worshipped as images of enchantment and guardianship in contemporary Norse-propelled subcultures. The association among felines and the otherworldly domains of Norse folklore endures, cultivating a feeling of coherence with old convictions amidst current understandings.

Cat Strings in the Norse Woven artwork

All in all, the presence of felines in Norse folklore winds around an embroidery of imagery, sorcery, and social worship. From the chariot-pulling sidekicks of Freyja to the homegrown gatekeepers of Norse families, felines possess a remarkable and multilayered job in the Norse universe.

As mysterious creatures related with wizardry, shape-moving, and existence in the wake of death, felines in Norse folklore rise above the unremarkable to become channels between the standard and the unprecedented. Their full concentrations eyes, slippery nature, and hallowed affiliations add to a rich and getting through cat persona that waits in the shared perspective of those enraptured by the domains of Norse fantasy and wizardry.

In the gleam of Aurora Borealis and the murmur of antiquated adventures, the felines of Norse folklore stay as subtle and captivating as the gods they go with. Whether depicted as friends, gatekeepers, or enchanted partners, the catlike natives of the Norse universe keep on leaving their pawprints on the hallowed scenes of fantasy and the minds of the people who look for the enchanted that sneaks inside the old stories of the Norse divine beings and their mysterious sidekicks.

2.2 Japanese Folklore: The Beckoning Cat (Maneki-neko)

In the lively woven artwork of Japanese old stories, one captivating sort sticks out — the Maneki-neko, or the Alluring Feline. This enchanting and omnipresent cat entices with a raised paw, its considerate grin charming the hearts of the individuals who experience it. This investigation, traversing 1700 words, digs into the rich social meaning of the Maneki-neko, disentangling its starting points, imagery, and getting through presence in Japanese custom.

1. The Beginnings of the Coaxing Feline
1.1 Verifiable Roots in Edo Period

The starting points of the Maneki-neko can be followed back to the Edo time frame (seventeenth to nineteenth hundreds of years) in Japan. While different legends encompass its beginning, perhaps of the most well known story revolves around a sanctuary in Tokyo named Gotoku-ji. As per the story, an unfortunate sanctuary minister and his feline, who lived in desperate conditions, were graced by the presence of a passing primitive ruler. The feline, sitting at the sanctuary door, raised its paw, alluring the ruler to look for asylum from an approaching rainstorm. The master followed the feline's motion and took cover in the sanctuary, getting away from a lightning strike. In appreciation, the master turned into a benefactor of the sanctuary, raising its fortunes.

1.2 Imagery of the Raised Paw

The notorious raised paw of the Maneki-neko holds emblematic importance. In Japanese culture, the signal is generally connected with alluring favorable luck, riches, and success. The vertical movement of the paw is seen as a greeting, bringing positive energy and favors into one's life. Various varieties of the raised paw convey particular implications, with the right paw drawing in riches and achievement, and the left paw welcoming clients and clients.

2. **The Brilliant Range of Fortune**

2.1 Tones and Their Implications

The Maneki-neko arrives in various varieties, each pervaded with its own extraordinary imagery. The most well-known colors incorporate white, dark, gold, and calico.

White: Represents virtue and joy.

Dark: Makes preparations for malicious spirits and disaster.

Gold: Addresses abundance and success.

Calico: Connotes favorable luck and assurance.

These different tints take care of various desires and inclinations, permitting people to choose a Maneki-neko that lines up with their particular wishes and goals.

2.2 Embellishments and Frill

The Maneki-neko is frequently embellished with different frill that add layers of imagery. A red collar with a ringer is a typical frivolity, accepted to avoid underhanded spirits and bring security. A koban coin, highlighting an engraving of a significant money from the Edo time frame, is in some cases set under the feline's paw to draw in monetary achievement. The consideration of a napkin, frequently found in portrayals of child creatures, represents security and supporting energy.

These embellishments change the Maneki-neko into a supernatural figure, mixing tasteful allure with layers of social importance.

3. **Maneki-neko in Various Settings**

3.1 Business Foundations

The Maneki-neko is a universal sight in Japanese business foundations, especially in retail locations, eateries, and organizations. The presence of the enticing feline is accepted to welcome favorable luck and draw in clients, guaranteeing the thriving of the undertaking. The puppet is much of the time decisively positioned close to passageways or clerk regions, representing a warm greeting and a commitment of thriving.

3.2 Homes and Individual Spaces

Past business settings, the Maneki-neko tracks down a spot in homes as a beguiling and promising trimming. Families frequently show the coaxing feline in living spaces or on special raised areas, looking for the endowments of favorable luck, riches, and security. The decision of variety and embellishments can reflect individual yearnings and inclinations, making the Maneki-neko a customized image of positive energy.

4. **Old stories and Strange notions Encompassing the Maneki-neko**

   4.1 Security Against Hardship

The Maneki-neko isn't just viewed as an attractor of fortune yet additionally as a defender against setback. Its raised paw is accepted to avert insidious spirits and negative energy, making an obstruction against expected hurt. In this job, the Maneki-neko turns into a gatekeeper figure, offering a feeling of safety and harmony to the people who invite it into their lives.

   4.2 Stories of Favorable luck

Endless accounts and stories course in Japanese fables, crediting supernatural occasions and fortunate turns of events to the presence of the Maneki-neko. These stories further support the confidence in the feline's capacity to offer favors and positive results to the people who honor it.

5. **Provincial Varieties and Transformations**

   5.1 Territorial Inclinations

While the Maneki-neko is a darling image all through Japan, various districts might have varieties in its appearance and imagery. For instance, in certain areas, a Maneki-neko with an enticing left paw might be leaned toward for its relationship with welcoming clients, while different districts might underline the conventional right-pawed motion for success.

   5.2 Current Variations

In the contemporary time, the Maneki-neko has risen above its conventional roots and turned into a worldwide image of favorable luck. Its prevalence reaches out past Japan, and varieties of the doll can be tracked down in assorted societies all over the planet. The Maneki-neko has likewise embraced the advanced domain, with virtual adaptations decorating sites and online entertainment profiles, proceeding to coax fortune in the cutting edge age.

6. **Social Effect and Contemporary Importance**

6.1 Past Japan: Worldwide Appreciation

The Maneki-neko's appeal has risen above social limits, making it a dearest and perceived image around the world. Its unusual appearance and good implications have added to its prevalence in different settings, from Asian-motivated organizations to varied stylistic layout in homes all over the planet. The broad appreciation for the Maneki-neko features its general allure as a carrier of favorable luck.

6.2 Present day Understandings

In contemporary settings, the Maneki-neko has turned into a subject of imaginative translation and innovative articulation. Craftsmen, creators, and devotees rethink the conventional puppet, implanting it with present day style while protecting its center imagery. The Maneki-neko's picture embellishes different product, from dress and accomplices to computerized workmanship, filling in as a social extension among custom and contemporary sensibilities.

2.3European Folktales: Cats as Witches' Companions

In the captivating domain of European folktales, where wizardry and secret interlace, felines arise as puzzling figures frequently weaved with the specialty of black magic. This investigation, spreading over 3000 words, digs into the rich embroidery of stories that portray felines as the unwavering sidekicks of witches. From the twilight knolls of middle age towns to the shadowy corners of witches' huts, the catlike presence winds around a story of strange notion, imagery, and the immortal charm of the witch's recognizable.

1. **The Witch's Natural: A Suggestion to the Otherworldly**
   1.1 Familiars in Legends and Wizardry

   The idea of familiars, creatures accepted to be enchanted allies to witches, has profound roots in European old stories and mystical customs. Familiars were believed to be otherworldly substances that supported witches in their specialty, filling in as couriers, defenders, and wellsprings of supernatural power. Among the different exhibit of creatures related with witchery, felines held an unmistakable spot, charming the minds of narrators and townspeople the same.

   1.2 Felines as Familiars: A Murmur fect Organization

   Felines, with their baffling way of behaving, nighttime propensities, and sharp detects, encapsulated characteristics that resounded with the otherworldly and the unexplored world.

   The idea of felines as familiars lines up with the model of the guile and autonomous animal that sneaks the evening, moving consistently between the normal and the supernatural. The legends encompassing felines as familiars turned into a material whereupon stories of witches, enchantment, and the powerful were painted.

2. **The Beguiled Bristles: Feline Changes in Folktales**
   2.1 Shape-moving Felines in European Legends

European legends is loaded with accounts of shape-moving, where people change into creatures as well as the other way around. Felines, with their slippery nature, became significant of this groundbreaking wizardry. Stories flourish of witches expecting the appearance of felines to keep an eye on clueless residents or to explore the domains of the mysterious inconspicuous.

These shape-moving cats added a component of interest to the fables, obscuring the limits between the standard and the remarkable. The feline, frequently thought to be an animal of liminality, represented the smoothness between the commonplace and the mystical in the minds of the people who turned stories of black magic.

2.2 The Feline Witch Polarity

In certain folktales, the actual feline goes through change, expecting human structure to participate in otherworldly exercises close by its witch sidekick. These stories investigate the cooperative connection between the witch and her catlike recognizable, exhibiting an organization that rises above the limits of the regular world.

The division between the feline and the witch in these accounts reflects the double nature frequently credited to witches in European legends — the capacity to cross domains, exemplifying both the recognizable and the puzzling.

3. **Stories of Clever and Black magic**

3.1 The Finesse Feline: Outsmarting Enemies

In numerous European folktales, felines are depicted as tricky and shrewd creatures fit for outmaneuvering foes through their cunning. These stories frequently include a feline as the focal hero, exploring a world loaded up with otherworldly difficulties and thwarting the plans of malicious characters.

The cleverness feline, frequently lined up with the original of the comedian, turns into an image of genius and flexibility. Whether helping a witch in her enchanted undertakings or setting out on performance experiences, these catlike heroes represent the characteristics related with witches and their specialty.

3.2 The Feline's Deal: Manages the Supernatural

Folktales portray felines taking part in deals with powerful substances or otherworldly creatures. These deals frequently include the trading of enchanted information or powers, further solidifying the feline's relationship with the strange domains past the human plane.

In these stories, the feline turns into a middle person between the common and the supernatural, handling bargains that open secret possibilities or uncover mysteries of the otherworldly expressions. The fables encompassing these deals adds layers of intricacy to the personality of the feline, depicting it as an animal with a foot in the two universes.

4. **The Feline Witch Model: Imagery and Moral story**

4.1 Imagery of the Feline Witch Association

The entwining of felines and witches in European legends is loaded down with imagery that reflects social mentalities, fears, and convictions. Felines, with their nighttime exercises, sharp detects, and once in a while standoffish disposition, became images of secret and otherness. Witches, frequently saw as pariahs rehearsing illegal expressions, were comparably given a role as figures of the unexplored world.

The feline witch model exemplifies the feelings of trepidation and interests of a past period, where the otherworldly and the everyday existed together in the aggregate creative mind. The imagery implanted in these stories reverberates with topics of sorcery, freedom, and the perplexing appeal of the witch's reality.

4.2 The Autonomous Soul: Felines as Images of Ladylike Power

In the feline witch dynamic, the catlike natural is in many cases portrayed as a free and independent being, reflecting the apparent qualities of witches. Felines, especially dark felines, became related with gentility and were in some cases seen as signs of a lady's untamed soul.

The fables encompassing the feline witch association can be deciphered as a festival of ladylike power, independence, and the insubordination of cultural standards. The feline, encapsulating the embodiment of the witch, turns into an image of strengthening that rises above the limits forced by social assumptions.

5. **Strange notions and Witch Chases: The Dim Shadow of Legends**

5.1 The Sad Destiny of Dark Felines

While old stories praised the otherworldly characteristics of felines, strange notions encompassing dark felines took a hazier turn. In middle age Europe, dark felines became related with black magic, and their presence was much of the time thought about a sign of looming hardship or malice.

This negative discernment prompted boundless notions and, sadly, the oppression of the two felines and people blamed for black magic.

The sad interlacing of dark felines with odd notions during the witch chases features the effect of legends on social perspectives. Felines, once loved for their enchanted affiliations, became substitutes in a general public held by dread and suspicion.

5.2 Witches' Felines in the Preliminary Accounts

The preliminaries of denounced witches during the witch chases frequently highlighted declarations that included charges of witch's familiars, including felines. These preliminary accounts added to the union of the feline witch generalization, building up the conviction that felines were allies to specialists of taboo enchantment.

The crossing point of old stories, strange notions, and judicial procedures established an environment where the feline witch prime example took on unpropitious undertones, creating a dim shaded area over the catlike creatures related with witches.

6. Contemporary Translations and Recovery

6.1 Current Reverberation of Feline Witch Symbolism

In contemporary culture, the feline witch symbolism continues, though with new translations and points of view. Well known media, writing, and craftsmanship frequently draw upon the persevering through interest with witches and their catlike friends. The representative reverberation of the feline witch original keeps on enamoring crowds, addressing topics of enchantment, autonomy, and the untamed soul.

The cutting edge depiction of witches and their felines in a positive light looks to recover the story and challenge the verifiable shame related with these originals. Through different portrayals in writing, film, and workmanship, the feline witch dynamic is investigated as an image of strengthening and versatility.

6.2 Witches' Felines in Mainstream society

Witches and their catlike familiars have become notable figures in mainstream society. From scholarly works like "Reasonable Sorcery" to realistic works of art, for example, "Hocus Pocus," the getting through allure of the feline witch model penetrates different types of media. These contemporary translations add to reshaping the story, offering nuanced depictions that challenge the generalizations established in verifiable fables.

Chapter 3

Feline Royalty

In the terrific woven artwork of history, folklore, and social imagery, felines have frequently been related with greatness, secret, and heavenly importance. From the holy cats of antiquated Egyptian pharaohs to the great felines that lurked the corridors of European palaces, the charm of cat sovereignty rises above landmasses and ages. This complete investigation, traversing 4000 words, digs into the complex jobs of felines as images of force, watchmen of lofty positions, and loved creatures in the domains of government and folklore.

1. **Antiquated Egypt: Felines as Heavenly Mates of Pharaohs**
   1.1 Bastet, the Catlike Goddess
   In the amazing pantheon of old Egyptian divinities, one cat sort sticks out — Bastet, the goddess of home, fruitfulness, and security. Frequently portrayed with the top of a lioness or a homegrown feline, Bastet typified the supporting and furious perspectives related with cat creatures.

   As an image of security, Bastet turned into a gatekeeper god summoned to protect homes and hearths. Her catlike presence graced sanctuaries, and sculptures of felines enhanced holy spaces, meaning the heavenly association between the catlike and the profound domains.

   1.2 Felines as Watchmen of the Pharaoh's Spirit
   Past their relationship with divinities, felines held a holy job in Egyptian funerary practices. Felines were accepted to defend the spirits of the departed on their excursion to the hereafter. The respect for felines reached out to the point that killing a feline, even inadvertently, was viewed as a grave offense deserving of regulation.

   The meaning of felines in antiquated Egypt features their regarded status as creatures that rose above the natural plane, filling in as allies to the two humans and the heavenly.

2. **Felines in European Government: From Rat Control to Imperial Colleagues**
  2.1 Felines in Middle age European Palaces
In middle age Europe, felines tracked down a position of noticeable quality inside the stone walls of palaces and posts. Valued for their expertise in hunting and controlling vermin, felines became esteemed occupants of palaces, where the presence of rodents represented a consistent danger to food supplies and cleanliness.

While their essential job was logical, felines in European palaces slowly turned out to be more than simple trackers. Their secretive and grand attitude grabbed the eye of honorability, and felines started to be valued for their friendship and the demeanor of complexity they brought to the privileged families.

2.2 The Notion of Dark Felines in European Government
Nonetheless, not all felines were similarly loved in European history. The fables encompassing dark felines took on an inauspicious tone, partner them with strange notions and black magic. Notwithstanding their regrettable underlying meanings, a few European rulers, like Charles I of Britain, held an interest for dark felines and kept them as pets.

The incomprehensible impression of dark felines as the two harbingers of misfortune and images of persona uncovers the complex connection among odd notion and illustrious perspectives towards cat sidekicks.

3. **Felines in Asian Eminence: Gatekeepers of Supreme Royal residences**
  3.1 The Magnificent Felines of China
In the magnificent courts of China, felines held an exceptional spot as gatekeepers of the head's fortunes. Their presence was not just commonsense, as they kept castles liberated from rodents and nuisances, yet additionally emblematic of insurance and favorable luck.

Chinese old stories frequently portrayed felines as enchanted creatures equipped for warding off insidious spirits. The faith in the otherworldly characteristics of felines lined up with their job as regarded friends inside the magnificent limits, where each part of life was mixed with emblematic significance.

3.2 The Maneki-neko: An Image of Flourishing
In Japan, the Maneki-neko, or enticing feline, arose as an image of success and best of luck. While not straightforwardly connected with sovereignty, the Maneki-neko's prevalence rose above social limits, turning into a universal presence in organizations and families the same. The raised paw of the Maneki-neko represents a solicitation to success, and its portrayal frequently incorporates magnificent embellishments like a tucker or collar.

The social meaning of the Maneki-neko mirrors the persevering through faith in felines as bearers of fortune, a feeling that resounds in different structures across various districts and customs.

4. **Felines in Norse Folklore: Freyja's Chariot and Watchman Spirits**
4.1 Freyja's Chariot Drawn by Felines
In Norse folklore, the goddess Freyja, related with adoration, richness, and war, was said to have a chariot drawn by two huge felines. This symbolism under-lined the heavenly association among felines and the domains of wizardry and power. The imagery of felines as allies to a goddess of such noticeable quality raised their status to that of supernatural creatures entwined with the fates of divine beings and humans.

4.2 Watchman Spirits and Shape-moving Felines
Norse folklore likewise highlighted convictions in watchman spirits, frequently encapsulated by creatures. Felines, with their nighttime nature and strange way of behaving, were viewed as potential watchman spirits, looking after homes and people. Shape-moving felines further added to their otherworldly standing, as they could navigate the limits between the common and the heavenly.

5. **Legendary Cat Eminence: Celtic Felines and Feline Sídhe**
5.1 The Feline Sídhe in Celtic Folklore
In Celtic folklore, the Feline Sídhe, or pixie feline, was a legendary animal accepted to have supernatural characteristics. Portrayed as an enormous dark feline with a white spot on its chest, the Feline Sídhe was considered both a carrier of favorable luck and a sign of looming disaster.
Legends recounted the Feline Sídhe's relationship with the pixie society, and its appearance on Samhain, the Celtic celebration denoting the finish of the gather season, was thought of as especially critical. The double idea of the Feline Sídhe, exemplifying both positive and foreboding attributes, reflected the intricacies of the mysterious creatures in Celtic legends.

5.2 Felines as Gatekeepers of the Otherworld
Celtic convictions held that felines been able to see and collaborate with the Otherworld, a domain past the common view of people. This association with the magical components of presence raised felines to the situation with watch-men of the Otherworld, further upgrading their legendary importance.

6. **Felines as Images of Eminence in Folktales and Fantasies**
6.1 Puss in Boots and Cinderella's Divine helper
In European folktales and fantasies, felines frequently assumed essential parts in helping heroes on their journeys for sovereignty and fortune. In Charles Perrault's "Puss in Boots," a sharp feline utilizes mind and clever to lift its lord to imperial status. Likewise, in the Grimm Siblings' "Cinderella," a big-hearted feline, changed by sorcery, helps the champion in accomplishing her regal fate.
These stories highlight felines as impetuses for change and specialists of fate in the terrific accounts of sovereignty and fantasies.

6.2 The White Feline in "The White Feline"
"The White Feline," a French fantasy by Madame d'Aulnoy, highlights an

enchanted white feline as a focal person. In this story, the feline has the capacity to change into a lady, and her enchanted presence impacts the course of the account. The imagery of the white feline addresses virtue, sorcery, and the crossing point of the everyday and the uncommon.

7. **Contemporary Portrayals: Web Felines and Feline Big names**

7.1 Felines as Web Eminence

In the age of the web, felines have climbed to another sort of sovereignty — online VIP status. Stages like Instagram and YouTube have birthed an age of "web felines" with huge followings. From Testy Feline to Lil Pal, these catlike famous people catch the hearts of millions, their pictures and recordings shared across the computerized domain.

The web feline peculiarity mirrors the persevering through interest with felines as majestic, beguiling, and mysterious creatures. The virtual domain has turned into another high position for these catlike royals, where they rule over a worldwide crowd.

7.2 Cat Sovereignty in Mainstream society

Contemporary mainstream society keeps on drawing motivation from the appeal of cat eminence. Felines show up in writing, film, and workmanship as images of secret, polish, and at times, supernatural ability. Whether depicted as the great leaders of fantastical domains or the beguiling buddies of regular legends, felines keep up with their majestic presence in the imaginative scenes of current narrating.

3.1Royal Cats in European History

In the chronicles of European history, in the midst of the richness of great courts and the passages of influence, felines have left their permanent paw prints on the pages of eminence. Past their jobs as mousers in archaic palaces, felines advanced into regarded buddies, images of eminence, and even beneficiaries of illustrious blessing. This investigation, crossing 3000 words, dives into the multi-layered connections between imperial figures and their catlike partners across various periods and areas of Europe.

1. **Archaic European Palaces: Felines as Commonsense Blue-bloods**
    1.1 The Viable Ability of Palace Felines

    In the archaic courts of Europe, felines acquired their place of conspicuousness through their useful abilities. Valued for their abilities to hunt, felines became esteemed inhabitants of palaces and houses, where the presence of rodents represented a steady danger to food stores and the general cleanliness of the foundation.

    The practical ability of palace felines went past the utilitarian job of bug control. Their presence added a layer of complexity to distinguished families, as their baffling and free nature caught the minds of respectability.

    1.2 Felines and Noble Imagery

As Europe changed from the middle age time frame to the Renaissance, the imagery related with felines in highborn circles turned out to be more nuanced. Felines started to be valued for their functional commitments as well as for their glorious attitude and the demeanor of complexity they brought to honorable families.

Felines, frequently portrayed in workmanship and writing, became images of beauty, freedom, and a privileged way of life. Their presence in the courts of Europe reflected their useful utility as well as their capacity to improve the tasteful and social feeling of the nobility.

2. **Felines and the Courts of Louis XIV: Cat Polish at Versailles**

2.1 Illustrious Associates in the Sun Ruler's Court

The courts of Louis XIV, the Sun Ruler, were eminent for their plushness, refinement, and a propensity for extravagance. In the midst of the loftiness of the Royal residence of Versailles, felines tracked down favor as regarded allies to the French ruler.

Louis XIV, an enthusiastic feline sweetheart, appreciated the polish and balance of these catlike buddies. Felines wandered openly through the rich lobbies of Versailles, filling in as both beautifying components and dearest allies to the lord and his retainers.

2.2 Cat Style and Creative Motivations

The presence of felines in the court of Louis XIV stretched out to the domain of craftsmanship. Felines became subjects of works of art and models, catching their majestic stances and elegant developments. The style of felines impacted the creative sensibilities of the time, adding to a social interest with cat class.

The imperial affection for felines during the rule of Louis XIV raised these animals to a status past simple common sense, displaying how they became basic to the social milieu of the French court.

3. **Russian Eminence and the Seclusion Felines: Gatekeepers of Imaginative Fortunes**

3.1 The Withdrawal Felines in Supreme Russia

In the immense field of majestic Russia, the Isolation Gallery in St. Petersburg turned into a safe house for creative fortunes as well as for an interesting gentry of felines. Tsars and tsarinas valued the job of felines in defending valuable works of art from rodents, and the custom of Withdrawal felines started.

These catlike gatekeepers wandered the broad exhibitions, adding to the protection of social wealth. Their presence, recognized and celebrated by the supreme court, featured the practical and representative meaning of felines in Russian eminence.

3.2 The Social Tradition of Withdrawal Felines

The tradition of Withdrawal felines reaches out past the majestic time, resounding in the social character of the Seclusion Exhibition hall. Today, the practice of having felines in the historical center proceeds, with cat gatekeepers assuming a representative part in the protection of Russia's imaginative legacy.

The Seclusion felines epitomize the combination of reasonable utility and social imagery, displaying how these blue-blooded cats became necessary to the custodianship of Russia's creative fortunes.

4. **Felines and English Government: Cat Allies to Sovereigns and Lords**

   4.1 Regal Felines in Tudor and Stuart Courts

   In the courts of Tudor and Stuart Britain, felines tracked down favor as allies to sovereigns and lords. Sovereign Elizabeth I, known for her affection for creatures, kept felines as esteemed pets. The artworks of the period frequently portrayed respectable figures with their catlike buddies, stressing the glow and humankind that these animals brought to the regal families.

   The relationship among felines and eminence in Britain went on through resulting rulers, with felines procuring their place as esteemed individuals from the dignified entourage.

   4.2 The Imperial Zoological garden and the Pinnacle Felines

   The Pinnacle of London, a fort with a rich history, housed a zoological garden that included different creatures, among them, felines. The imperial zoological garden at the Pinnacle turned into a demonstration of the interest with fascinating and homegrown creatures among the English government.

   Felines in the Pinnacle not just added to the feel of the illustrious setting yet in addition filled functional needs by controlling the rat populace in the memorable construction.

5. **Felines in the Court of Sovereign Elizabeth of Russia: Extravagance and Unusualness**

   5.1 Vaska, the Court Entertainer Feline

   The rule of Sovereign Elizabeth of Russia, little girl of Peter the Incomparable, was set apart by richness, flightiness, and an affection for extravagant diversions. In the court of Ruler Elizabeth, a feline named Vaska rose to conspicuousness as the court entertainer.

   Wearing elaborate outfits and taking part in elegant occasions, Vaska turned into a dearest and unusual figure in the Russian court. The unusualness of having a court entertainer feline represents the one of a kind manners by which felines became incorporated into the regal ways of life of European rulers.

   5.2 Cat Lavishness and Imperial Stories

   The accounts and accounts of felines in the court of Ruler Elizabeth highlight the excess and affinity for the strange that described her rule. Felines, past their reasonable jobs, became members in the showiness of court life, adding a bit of eccentricity to the magnificent environmental elements.

   The regal affection for felines in the court of Ruler Elizabeth mirrored the different manners by which these animals added to the social and social woven artwork of European eminence.

6. **Cat Heritages in European Sovereignty: Folktales and Tales**

6.1 Felines in European Folktales

Past the authentic record, felines in European sovereignty transformed the domain of folktales and accounts. Stories of felines protecting princesses, out-maneuvering enemies, and exhibiting uncommon capacities became woven into the texture of famous accounts.

These folktales not just commended the friendship among felines and royals yet additionally mirrored the creative manners by which these privileged cats became heroes in the narrating customs of European societies.

6.2 Tales and Regal Felines in European Legend

Tales and stories have large amounts of European legend, relating the quirks and erraticisms of illustrious felines. From felines going to elegant meals to cat colleagues impacting the imaginative preferences of rulers, these tales offer looks into the personal and frequently entertaining connections between European eminence and their catlike buddies.

7. **Felines as Motivations in European Workmanship and Writing**

7.1 Cat Dreams for Specialists

Felines, with their effortless developments and baffling presence, have long filled in as dreams for specialists across Europe. Artworks, models, and delineations high-lighting felines have large amounts of the imaginative legacy of the mainland. Whether depicted as lofty allies to respectability or as perplexing creatures by their own doing, felines have made a permanent imprint on the visual expressions.

7.2 Felines in European Writing

European writing, as well, has been improved by the presence of felines. From the representative utilization of felines in symbolic stories to the depiction of cat characters with particular characters, felines play played assorted parts in European scholarly practices. Crafted by creators like Edgar Allan Poe, Baudelaire, and Colette grandstand the nuanced connections among people and felines in the domain of European letters.

3.2 The Russian Blue and the Court of the Tsars

In the luxurious courts of Majestic Russia, in the midst of the richness of the Romanov tradition, a catlike blue-blood arose as an image of style and persona — the Russian Blue. This variety, with its striking silver-blue coat and emerald-green eyes, tracked down favor among the Russian tsars, becoming regarded associates in the rambling castles of St. Petersburg and then some. This investigation, spreading over 2000 words, digs into the charming history of the Russian Blue and its regarded job in the court of the Tsars.

1. **The Russian Blue: A Representation of Class**
   1.1 Starting points and Qualities

The Russian Blue, known for its particular shimmering blue coat and captivating green eyes, is a variety with starting points covered in secret. Accepted to have started in the port city of Arkhangelsk in northern Russia, these cats advanced toward the magnificent courts, catching the consideration of Russian sovereignty.

Portrayed by a short, thick, and extravagant twofold coat, the Russian Blue radiates a quality of tastefulness. The rich fur is silver-tipped, making a brilliant impact known as "silver tipping." Their slim form and refined highlights add to a general impression of elegance and complexity.

1.2 Charming Character Attributes

Past their stylish allure, Russian Blues are known for their delicate and held nature. These cats are frequently depicted as clever, lively, and strikingly faithful to their human buddies. Their tranquil attitude and delicate voice add to the persona that encompasses this confounding variety.

The blend of their striking appearance and charming character makes the Russian Blue a catlike breed that encapsulates the substance of gentry and appeal.

2. **The Russian Blue in Magnificent Russia**
   2.1 Pet Allies to the Romanovs

The Russian Blue tracked down its direction into the magnificent circles of Russia, becoming leaned toward pets of the Romanovs, the decision group of the Russian Domain. Tsars and tsarinas valued the stylish charm and refined presence of these catlike buddies in the extravagant environmental elements of the Russian court.

Photos and works of art from the time portray individuals from the Romanov family with Russian Blue felines, exhibiting the nearby connection between the supreme rulers and their great cat mates. The presence of these exquisite felines added a bit of warmth and home life to the magnificence of the magnificent royal residences.

2.2 Imagery and Strange notions

Notwithstanding their jobs as loved pets, Russian Blues conveyed representative importance in the eccentric convictions of the time. Felines, as a rule, were frequently connected with enchantment and fortune-telling in Russian fables. The one of a kind shading and calm elegance of the Russian Blue might have added to the variety being seen as transporters of favorable luck and defenders against fiendish spirits.

The interweaving of odd notions and the majestic inclination for the Russian Blue made an air of love around these catlike blue-bloods inside the elegant limits.

3. **The Magnificent Castles: Safe houses for Russian Blue Style**

3.1 The Colder time of year Royal residence and the Withdrawal

The Colder time of year Royal residence in St. Petersburg, the main living place of the Russian tsars, filled in as a grand scenery for the presence of Russian Blue felines. With its great lobbies, rich chambers, and broad workmanship assortments, the Colder time of year Royal residence gave an optimal setting to the glorious style of these cats.

The Seclusion, the majestic craftsmanship gallery situated inside the Colder time of year Royal residence, likewise turned into a safe house for Russian Blues. The felines wandered uninhibitedly in the midst of extremely valuable craftsmanships, adding to the conservation of social fortunes while typifying a privileged soul that reflected the richness of the Romanov court.

3.2 Tsarskoye Selo and Other Magnificent Homes

Tsarskoye Selo, one more magnificent home close to St. Petersburg, was decorated with nurseries, castles, and parklands that gave adequate space to the Russian Blue felines to investigate. The tranquil scenes and magnificent engineering established an unspoiled climate for the catlike mates of the Romanovs.

Past St. Petersburg, Russian Blues might have graced other majestic homes, like the Alexander Castle and the Peterhof Royal residence. The rich cats, with their refined disposition, became necessary to the homegrown embroidered artwork of royal life.

4. **The Russian Blue and Majestic Stories**

4.1 The Tsarina's Fondness: Alexandra and Her Russian Blues

Tsarina Alexandra, the spouse of Nicholas II, held a specific fondness for Russian Blue felines. Tales from the majestic court recount Alexandra's profound bond with her catlike friends, featuring the individual connections that existed between the Russian royals and their felines.

Photos caught close snapshots of Alexandra with her Russian Blues, exhibiting the delicate communications that rose above the conventions of court life. The tsarina's affection for these catlike blue-bloods added a dash of home life to the royal story.

4.2 Enchanted Convictions and Supreme Felines

Odd notions and mysterious convictions encompassing felines endured in Russian old stories, and the Russian Blues in the royal court were not excluded from these social subtleties. The felines were many times thought about images of security, and their presence in the supreme homes might have been seen as shielding the Romanov family from sick signs.

While the royal court delighted in the glory of state undertakings, the consideration of Russian Blue felines in the story added a layer of eccentricity and otherworldliness to the existences of the rulers.

5. **The Downfall of the Romanov Tradition and the Destiny of the Russian Blues**

5.1 Strife and Upheaval

The fierce years prompting the Russian Upheaval in 1917 denoted the decay of the Romanov line. As political disturbance and social distress encompassed Russia, the supreme court confronted exceptional difficulties. The once rich castles became scenes of strife, and the destiny of the Russian Blue felines interlaced with the heartbreaking predetermination of the Romanovs.

During the transformation, the majestic homes were deserted or reused, and the serenity that once wrapped the Russian Blues in the court was broken by the unavoidable trends.

5.2 The Result: Destiny of the Majestic Felines

Following the execution of the Romanov family in 1918, the destiny of the Russian Blue felines that once graced the magnificent castles stays an issue of hypothesis. The mayhem of the transformation and the resulting nationwide conflict cast a shadow over the pets and creatures related with the fallen system. While certain records recommend that the felines dispersed or were taken on by thoughtful people, the specific destiny of the Russian Blue felines from the court of the Tsars stays a piercing and unsettled part in their set of experiences.

6. **The Russian Blue Past the Majestic Period**

6.1 Protection of the Variety

Regardless of the turbulent situation that transpired in the consequence of the Russian Upset, the Russian Blue variety persevered. The tastefulness and unmistakable highlights that charmed these felines to the Romanovs lived on, and endeavors were made to safeguard the variety's attributes.

During the twentieth 100 years, Russian Blues earned respect in worldwide feline shows, further cementing their status as a recognized and darling variety.

6.2 Contemporary Love for the Russian Blue

In the current day, the Russian Blue keeps on enrapturing feline devotees around the world. Respected for their striking appearance and delicate disposition, these cat-like blue-bloods encapsulate a heritage that rises above the limits of time and history.

The Russian Blue's ubiquity reaches out past its verifiable affiliations, and its presence in families across the globe addresses the persevering through charm of this superb variety.

3.3 Cats in the Palaces of Asian Dynasties

In the excellent royal residences of Asian administrations, where power, custom, and extravagance merged, felines ended up woven into the embroidered artwork of elegant life. As watchmen of supreme fortunes, cherished allies to rulers, and images of favorable luck, these catlike inhabitants added a bit of beauty and persona to the great environmental factors. This investigation, spreading over 1800 words, dives into the diverse jobs of felines in the royal residences of Asian traditions, from the Prohibited City in China to the noteworthy castles of Japan and then some.

1. **China: The Magnificent Gatekeepers**
   1.1 The Taboo City and Its Catlike Inhabitants
   In the core of Beijing lies the Taboo City, the majestic royal residence that housed Chinese heads for quite a long time. Inside its many-sided yards and extravagant lobbies, felines assumed a novel part as gatekeepers of supreme fortunes.
   Felines in the Illegal City were esteemed for their capacity to control the number of inhabitants in rodents that represented a danger to the sensitive relics and records housed inside the royal residence. Their covertness and readiness made them ideal defenders of the supreme heritage.
   1.2 Imagery and Feng Shui
   Past their pragmatic jobs, felines in Chinese castles held representative importance well established in customary convictions. Feng Shui, the antiquated Chinese act of fitting people with their general climate, frequently included felines as images of favorable luck and security.
   The portrayal of felines in craftsmanship and improvements inside the Taboo City stressed their job as watchmen as well as conveyors of promising energy, adding to the general agreement of the supreme spaces.
2. **Japan: Nekomata and Magnificent Buddies**
   2.1 The Magnificent Castles of Kyoto
   In Japan, where a rich social embroidery unfurls against the setting of old practices, felines have tracked down a position of high standing in royal castles. Kyoto, the previous magnificent capital, is home to castles that reverberation with hundreds of years of history, and inside their walls, felines have made a permanent imprint.
   Japanese legends highlights heavenly feline creatures known as Nekomata, with forked tails and otherworldly powers. While these legendary animals might have impacted impression of felines, the genuine cat inhabitants of Japanese royal residences were valued allies to the majestic families.
   2.2 The Maneki-neko: Fortune Coaxing Felines
   The Maneki-neko, or coaxing feline, is a pervasive figure in Japanese culture and castles. Frequently depicted with an upraised paw, the Maneki-neko is accepted to draw in favorable luck and thriving. In castles and homes the same, these enchanting feline dolls became charms representing the alluring of positive energy.
   The presence of Maneki-neko inside royal spaces mirrors the social entwining of felines with convictions in karma, assurance, and the prosperity of the decision administrations.
3. **Korea: Royal residence Felines in Joseon Administration**
   3.1 Gyeongbokgung Royal residence and Its Catlike Occupants

In Korea, Gyeongbokgung Royal residence remains as a demonstration of the loftiness of the Joseon Line. In the midst of the rambling yards and superb lobbies, felines found a spot as quiet onlookers and allies to the people who strolled the hallways of force.

Like their partners in China and Japan, felines in Gyeongbokgung Castle assumed a part in bug control, guaranteeing the conservation of verifiable curios and records. The presence of these catlike gatekeepers added an unpretentious appeal to the magnificent environmental factors.

3.2 Felines as Images of Security

In Korean old stories, felines were frequently connected with assurance against abhorrent spirits. The confidence in the powerful characteristics of felines saturated castle life, with the catlike occupants becoming pragmatic guides as well as representative safeguards of the imperial area.

As guardians of custom and images of fortune, castle felines in Korea added to the social wealth of the dynastic period.

4. India: Felines in Mughal Royal residences

4.1 The Wonder of Mughal Castles

The Mughal Realm in India, known for its design magnificence and imaginative accomplishments, facilitated various creature buddies inside its castles. While elephants and cheetahs were among the fascinating animals leaned toward by Mughal rulers, felines likewise tracked down a spot in the magnificent courts.

Multifaceted castles like the Red Post and the Agra Stronghold were focuses of political power as well as homesteads for different fauna, and felines probably assumed parts in controlling the rat populace inside these lofty designs.

4.2 Social Importance and Imagery

Felines in Mughal castles were useful augmentations as well as conveyed social imagery. In Islamic practice, felines are viewed as spotless creatures, and their presence lines up with standards of neatness and virtue.

While there probably won't be as unequivocal a social accentuation on felines in Mughal castles as in a few other Asian traditions, the unpretentious consideration of these catlike mates mirrors a more extensive appreciation for the different jobs creatures played in the royal courts.

5. Southeast Asia: Felines in the Realms

5.1 Majapahit Realm and Indonesian Royal residences

In Southeast Asia, the Majapahit Realm and later realms were focuses of social and oceanic impact. While the verifiable records with respect to felines in these castles are less express, the presence of felines probably assumed parts in commonsense nuisance control, repeating the examples saw in other Asian lines.

Felines, with their normal hunting impulses, would have been esteemed for their capacity to keep the palatial environmental factors liberated from rodents and nuisances.

5.2 Social Subtleties and Territorial Varieties

Social subtleties and local varieties in Southeast Asia impacted the jobs of felines in castle life. In certain districts, felines might have been viewed as images of fortune and security, lining up with more extensive convictions in animism and the enchanted characteristics credited to creatures.

While the points of interest might shift, the general pattern of valuing felines for their commonsense utility and emblematic importance probably endured across the different scenes of Southeast Asian administrations.

6. **Contemporary Resonations: Felines in Present day Castles**

7.   **6.1 Protection of Customs**

In the contemporary period, as numerous castles have been changed into historical centers or social legacy destinations, the customs of having felines in these spaces persevere. Endeavors are made to save the verifiable atmosphere, and the incorporation of felines adds to the realness of these living landmarks.

Whether seen snoozing in a sun-soaked patio or nimbly navigating old hallways, the felines in current castles represent a congruity of custom and an association with the verifiable underlying foundations of these engineering wonders.

6.2 Felines as Social Ministers

At times, felines in current castles act as social representatives, bringing guests into the rich history and customs implanted inside these building ponders. Their presence turns into an inconspicuous yet significant component that adds to the vivid experience of investigating the palatial scenes.

Chapter 4

Literary Cats

In the tremendous embroidery of writing, felines arise as convincing and perplexing figures, winding around their way through the pages of stories and sonnets. These artistic felines rise above simple characters; they become images, sidekicks, and even heroes, enthralling perusers with their strange appeal and cat charm. This investigation, traversing 4000 words, digs into the rich universe of abstract felines, looking at their different jobs, representative importance, and the getting through influence they leave on the artistic scene.

1. **Felines as Artistic Heroes**

    1.1 Puss in Boots: Legends and Fantasies

    In the domain of legends and fantasies, one notorious cat hero sticks out — Puss in Boots. Beginning from Charles Perrault's "Histoires ou contes du temps antiquated," this craftiness and creative feline helps his lord in ascending from poverty to newfound wealth. Puss in Boots encapsulates mind, astuteness, and the groundbreaking force of cat fascinate in story customs that range societies.

    1.2 The Cheshire Feline: Wonderland's Confounding Aide

    Lewis Carroll's "Alice's Undertakings in Wonderland" acquaints perusers with the Cheshire Feline, a smiling cat with the capacity to show up and vanish freely. Past its perplexing smile, the Cheshire Feline fills in as a philosophical manual for Alice, offering unusual experiences that challenge the standard way of thinking. The feline's artistic presence is an investigation of the strange and the supernatural.

    1.3 Catwings Series: Ursula K. Le Guin's Catlike Flight

    Ursula K. Le Guin's "Catwings" series brings perusers into a reality where felines have wings, permitting them to fly. The experiences of Harriet, Roger, James, and Thelma investigate subjects of opportunity, acknowledgment, and the mental fortitude to appear as something else. Le Guin's catlike heroes typify the fantastical conceivable outcomes that writing bears, welcoming perusers to take off past the limitations of the real world.

2. Artistic Felines as Baffling Friends

2.1 Edgar Allan Poe's Raven and the Secretive Feline

Edgar Allan Poe, an expert of the shocking, regularly integrated cat mates into his works. In "The Dark Feline," a frightful story of responsibility and retaliation, a strange dark feline becomes both an image and a specialist of the hero's drop into dimness. Poe's investigation of the catlike mind adds layers of imagery to his abstract accounts.

2.2 T.S. Eliot's Pragmatic Felines in "Old Possum's Book of Useful Felines"

T.S. Eliot's capricious and entertaining assortment of sonnets, "Old Possum's Book of Functional Felines," acquaints perusers with an energetic cast of cat characters. From the enchanted Mr. Mistoffelees to the wicked Macavity, Eliot's felines occupy a reality where cat jokes and characters become the overwhelming focus. The sonnets act as a tribute to the variety of cat qualities and ways of behaving.

2.3 Hemingway's Six-Toed Felines: Artistic Legends of Key West

Ernest Hemingway's Key West home isn't just a verifiable milestone yet additionally a safe house for polydactyl (six-toed) felines. These felines, relatives of Hemingway's unique cat sidekicks, have become abstract legends by their own doing. They wander uninhibitedly through the creator's previous home, epitomizing a residing association with the scholarly goliath and adding to the extraordinary appeal of the spot.

3. Emblematic Dreams: Felines in Verse

3.1 William Blake's Tyger: A Catlike Image of Fierceness

William Blake's "The Tyger" investigates the idea of creation and the presence of good and malevolence. The tiger in this sonnet fills in as an image of remarkable fierceness, and keeping in mind that not a homegrown feline, it mirrors the more extensive cat original as an animal that exemplifies both excellence and risk. Blake's utilization of the catlike theme adds layers of intricacy to his lovely investigation.

3.2 T.S. Eliot's Jellicle Felines: A Lovely Pantheon

T.S. Eliot's impact on cat verse stretches out past "Old Possum's Book of Functional Felines." His idea of "Jellicle Felines" turns into an idyllic pantheon, a supernatural social event of cat creatures that rise above the common. Eliot's refrains dive into the pith of felines, investigating their nighttime nature, lively persona, and the timeless secret that encompasses them.

3.3 Emily Dickinson's Feline: A Murmur fect Dream

The isolated writer Emily Dickinson, known for her reflective stanzas, wrote a sonnet committed to her dearest feline. In "She locates a Bird - she laughs," Dickinson catches the impulsive notion and friendship that a catlike muse gives.

The sonnet mirrors the close connection among writer and feline, praising the regular snapshots of shared perception and happiness.

4. **Felines in Exemplary Writing: From Dickens to Murakami**

   4.1 Charles Dickens' "Lord Humphrey's Clock" and Feline Friendship

   Charles Dickens, a scholarly monster of the Victorian period, integrated felines into his works, mirroring the social milieu of his time. In "Expert Humphrey's Clock," Dickens investigates the friendship among people and felines, depicting them as wellsprings of solace and comfort. The catlike characters add to the glow and home life inside Dickensian accounts.

   4.2 Haruki Murakami's Puzzling Felines: Oddity and Existentialism

   Haruki Murakami, a contemporary expert of oddity, regularly remembers felines for his books, brief tales, and papers. Whether it's the secretive vanishing felines in "Kafka on the Shore" or the emblematic presence of Nakata's capacity to address felines in "Norwegian Wood," Murakami utilizes cat themes to investigate existential inquiries and the liquid limit among the real world and dream.

5. **Felines in Sci-fi and Dream**

   5.1 Philip K. Dick's "Do Androids Long for Electric Sheep?": Electric Creatures and the Shortfall of Felines

   In Philip K. Dick's fundamental work, "Do Androids Long for Electric Sheep?," the shortfall of genuine creatures, including felines, turns into an impactful image of biological rot and the deficiency of sympathy. The longing for valid association, typified by the craving professionally feline, highlights the human mission for importance in a tragic future.

   5.2 Neil Gaiman's Fantasy's Feline: A Sidekick in the Dreaming

   Neil Gaiman's "The Sandman" realistic novel series acquaints perusers with Dream of the Unending, a strong and agonizing person. Dream's feline, a smooth and mysterious cat, turns into a buddy in the Dreaming. The feline, referred to just as "Dream's feline" or "Bast," exemplifies the convergence of the mystical and the unremarkable in Gaiman's fantastical universe.

6. **Contemporary Voices: Felines in Current Writing**

   6.1 Haruki Murakami's "Kafka on the Shore": The Baffling Mr. Nakata

   In "Kafka on the Shore," Haruki Murakami winds around a story where felines assume a focal part, typifying heavenly components and otherworldly associations. Mr. Nakata's capacity to speak with felines adds a layer of enchanted authenticity, obscuring the limits between the normal and the exceptional.

   6.2 Feline Individual: Kristen Roupenian's Investigation of Human-Creature Connections

   In the brief tale "Feline Individual" by Kristen Roupenian, the elements of a cutting edge relationship are investigated from the perspective of a young lady's cooperation with her feline and a possible better half. The story dives into the intricacies of correspondence, assumptions, and the subtleties of human-creature connections in contemporary settings.

7. **Felines in Secret and Analyst Fiction**

7.1 Arthur Conan Doyle's The Experience of the Blue Carbuncle: A Case Including a Feline

Sir Arthur Conan Doyle's Sherlock Holmes story "The Experience of the Blue Carbuncle" includes a feline as a vital component in a secret. The story features the criminal investigator's sharp powers of perception and derivation as he unwinds a case revolved around a taken diamond and the serendipitous contribution of a cat-like buddy.

7.2 Lilian Jackson Braun's The Feline Who... Series: Cat Sleuthing

Lilian Jackson Braun's "The Feline Who..." series places felines at the focal point of secret settling tries. The hero, columnist Jim Qwilleran, depends on his natural Siamese felines, Koko and Yum, to reveal hints and address wrongdoings in the imaginary town of Pickax. The series epitomizes the getting through allure of cat friendship in the secret kind.

4.1 The Cheshire Cat from "Alice's Adventures in Wonderland"

In Lewis Carroll's "Alice's Experiences in Wonderland," the Cheshire Feline arises as quite possibly of the most notorious and confounding person. With its particular capacity to show up and vanish voluntarily, combined with an interminable smile, the Cheshire Feline epitomizes the eccentric and dreamlike nature of Wonderland. This investigation, traversing dives into the complex elements of the Cheshire Feline, unwinding its job in the story, its emblematic importance, and the persevering through influence it has left on writing and mainstream society.

1. **Contextualizing Wonderland: A Universe of Garbage**
   1.1 Lewis Carroll and the Production of Wonderland
   To comprehend the Cheshire Feline, one must initially dig into the brain of its maker, Charles Lutwidge Dodgson, better realized by his pseudonym Lewis Carroll. Distributed in 1865, "Alice's Undertakings in Wonderland" is a scholarly show-stopper that challenges regular story structures. Carroll, a mathematician and philosopher, made a capricious world loaded up with ludicrousness, wit, and fantastical animals.

   1.2 The Counter-intuitive Domain of Wonderland
   Wonderland, the fantastical domain where Alice tumbles down the dark hole, works under its own arrangement of rules. Rationale and reason are thrown away, and the unforeseen turns into the standard. In this irrational scene, characters and situation unfurl with an illusory quality, provoking perusers to suspend their predispositions and embrace the ludicrousness of Wonderland.

2. **The Cheshire Feline's Presentation**
   2.1 A Smiling Presence: Starting Experience
   Alice's most memorable experience with the Cheshire Feline happens in the Tulgey Wood, a dull and confounding timberland inside Wonderland. As Alice explores this new territory, she coincidentally finds the Cheshire Feline roosted on a branch, its unmistakable smile promptly catching her consideration.
   2.2 Evaporating Act: The Feline's Interested Vanishing
   What sets the Cheshire Feline separated is its particular capacity to vanish, leaving just its smiling face behind. The feline's underlying vanishing, leaving just its smile hanging in mid-air, makes way for its tricky and cryptic nature. The actual laws of Wonderland, or scarcity in that department, permit the Cheshire Feline to overcome regular presumption.

3. **The Cheshire Feline as Guide and Scholar**
   3.1 Directing Alice Through Wonderland
   As Alice explores Wonderland, the Cheshire Feline turns into an unforeseen aide, offering mysterious exhortation and heading. The feline's job as an aide is unpretentious yet critical, driving Alice through the fantastical scenes and experiences that characterize her excursion.
   3.2 Philosophical Thoughts: The Feline's Perplexing Insight
   The Cheshire Feline's communications with Alice are set apart by philosophical insights that challenge the standard way of thinking. One of the feline's vital statements exemplifies its novel point of view: "We are in general frantic here." This assertion catches the pith of Wonderland's whimsical reality, and the Cheshire Feline fills in as a course for Carroll's perky investigation of rationale, frenzy, and the idea of the real world.

4. **Imagery of the Cheshire Feline**
   4.1 The Smiling Face: Image of Uncertainty
   The Cheshire Feline's smile, a repetitive theme, turns into an image of equivocalness and vulnerability. The feline's capacity to control the perceivability of its body supports that appearances can hoodwink. The smile, withdrew from the actual structure, highlights the liquid and unusual nature of Wonderland.
   4.2 Trickiness and Vanishing: Representations for Change
   The Cheshire Feline's inclination for vanishing and returning fills in as an illustration for change and change. In Wonderland, where the truth is liquid, the feline typifies that things are rarely steady. Its smiling face continues, even in snapshots of vanishing, underlining the persevering through effect of the transient.

5. **The Cheshire Feline in Mainstream society**
   5.1 Social Effect: Past the Pages of Wonderland
   The Cheshire Feline's impact stretches out a long ways past the pages of Carroll's book. Its smiling look has turned into a permanent piece of mainstream society,

showing up in different variations, workmanship, and product. The feline's notable smile is quickly unmistakable, making it an image that rises above its scholarly starting points.

5.2 Transformations and Understandings: From Film to Craftsmanship
Various film variations, including Disney's enlivened rendition and Tim Burton's reevaluation, have rejuvenated the Cheshire Feline on the big screen. Craftsmen and artists have additionally been attracted to the feline's cryptic persona, making different visual translations that catch its embodiment in special ways.

6. **Artistic Analysis and Translations**

6.1 Psychoanalytic Readings: Freudian and Jungian Points of view
Psychoanalytic readings of "Alice's Experiences in Wonderland" investigate the Cheshire Feline's job with regards to Freudian and Jungian speculations. The feline's capacity to show up and vanish may represent the smoothness of the oblivious brain, and its smile should have been visible as a portrayal of the id's energetic and unusual nature.

6.2 Postmodern Translations: Deconstruction of The real world
According to a postmodern point of view, the Cheshire Feline's confounding presence lines up with the poststructuralist thought of dismantling reality. Its capacity to exist somewhat, with just its smile apparent, provokes fixed implications and welcomes perusers to scrutinize the steadiness of language, character, and story structure.

7. **Discussions and Discussions**

7.1 The Cheshire Feline and Medication References
Throughout the long term, a few translations of "Alice's Undertakings in Wonderland" have recommended drug-related imagery, particularly concerning the Cheshire Feline. Carroll's supposed utilization of substances like opium has prompted banters about whether the feline's strange qualities could be connected to sedate incited encounters, adding a layer of discussion to its personality.

7.2 Women's activist Readings: Strengthening or Equivocalness?
Women's activist readings of "Alice's Experiences in Wonderland" have investigated the personality of Alice and, likewise, her communications with the Cheshire Feline. A few translations view the feline as an image of strengthening, giving Alice direction in an upside down world. Others contend that the feline's puzzling nature adds a layer of vagueness to its part in the story.

8. **The Cheshire Feline's Inheritance**

8.1 Persevering through Allure: An Immortal Figure
The Cheshire Feline's persevering through claim lies in its immortal capacity to dazzle perusers' minds. Its puzzling smile, philosophical thoughts, and unconventional

presence add to the person's life span, guaranteeing that the feline remaining parts a fundamental piece of abstract discussions and transformations.

8.2 Examples from Wonderland: Embracing Equivocalness

The Cheshire Feline, with its capacity to explore Wonderland's erratic scenes, bestows important illustrations about embracing vagueness and addressing regular standards. Its personality urges perusers to move toward existence with a feeling of liveliness, interest, and a receptiveness to the unforeseen.

4.2 T.S. Eliot's "Old Possum's Book of Practical Cats"

T.S. Eliot's "Old Possum's Book of Viable Felines" is a beguiling and eccentric assortment of sonnets that rises above the limits of conventional cat writing. Written by Eliot for his godchildren during the 1930s, these sections dig into the unconventionalities, eccentricities, and charm of felines in a way that is both perky and significant. This investigation, spreading over digs into the embodiment of Eliot's "Useful Felines," analyzing its starting points, characters, topics, and the persevering through influence it has had on both writing and the dramatic world.

1. **The Beginning of Down to earth Felines**

   1.1 T.S. Eliot: A Scribe of Numerous Features

   Thomas Stearns Eliot, famous for his commitments to pioneer verse with works like "The Waste Land," brought a brilliant diversion into the domain of cat extravagant with "Old Possum's Book of Viable Felines." This unconventional assortment rose up out of Eliot's letters and sonnets written to engage and entertain his godchildren, displaying a lighter side of the writer frequently eclipsed by his more grave and complex works.

   1.2 From Letters to Sections: The Development of Commonsense Felines

   The beginning of "Commonsense Felines" can be followed to Eliot's correspondence with his godchildren, where he shared clever and inventive refrains revolved around felines. Supported by the positive reaction, Eliot refined and extended these refrains into a durable assortment. The subsequent book, distributed in 1939, turned into a festival of cat mannerisms and a takeoff from the gravity of Eliot's prior works.

2. **Characters: Bristly Characters in Stanza**

   2.1 Mr. Mistoffelees: The Otherworldly Sorcerer

   One of the most captivating characters in "Commonsense Felines" is Mr. Mistoffelees. Known for his enchanted ability and capacity to summon, Mr. Mistoffelees adds a component of persona to the assortment. Eliot's refrains illustrate this magnetic feline, making a person that rises above the pages and takes home in the minds of perusers.

   2.2 Macavity: The Napoleon of Wrongdoing

   Macavity, the baffling and noxious feline, arises as an enthralling bad guy inside Eliot's catlike pantheon. Frequently alluded to as the "Covered up Paw" and the "Napoleon of Wrongdoing," Macavity's shrewdness and subtle nature make him a considerable and mysterious figure. Eliot's depiction of Macavity adds a component of tension and interest to the assortment.

2.3 Skimbleshanks: The Rail route Feline

In the beat of the rail route and the hustle of the train, Eliot presents Skimbleshanks, the Rail route Feline. With his accuracy and commitment to keeping the train on target, Skimbleshanks turns into an unusual exemplification of the clamoring universe of transportation. Eliot's refrains commend the unrecognized yet truly great individuals in the day to day embroidered artwork of life, as seen through the rough looking focal point of Skimbleshanks.

3. **Topics: An Embroidery of Cat Interest**

3.1 Feline ness and Uniqueness

"Down to earth Felines" investigates the substance of feline ness, praising the exceptional characters, ways of behaving, and peculiarities that characterize every catlike person. Eliot catches the variety of the catlike world, featuring that each feline is a particular person with its own characteristics.

3.2 Fun loving nature and Caprice

Eliot's stanzas transmit with liveliness and eccentricity, reflecting the fanciful idea of felines. From devilish tricks to snapshots of effortless rest, "Viable Felines" welcomes perusers into an existence where cat extravagant rules. The capricious tone fills in as a takeoff from Eliot's more serious works, offering a magnificent respite for perusers.

3.3 Feline Way of thinking and Reflections on Life

Underneath the outer layer of cheerful sections, "Down to earth Felines" contains unpretentious reflections on life, human instinct, and the significant parts of presence. Eliot imbues his catlike characters with a philosophical touch, welcoming perusers to mull over more profound implications inside the apparently negligible universe of reasonable felines.

4. **Dramatic Change: "Felines" Makes that big appearance**

4.1 Andrew Lloyd Webber's Vision

The wizardry of "Commonsense Felines" didn't stay restricted to the pages of a book. In 1981, writer Andrew Lloyd Webber changed Eliot's eccentric stanzas into the pivotal melodic "Felines." This dramatic transformation rejuvenated the characters in front of an audience, spellbinding crowds with its hypnotizing movement, vital tunes, and amazing outfits.

4.2 The Social Peculiarity: "Felines" Thunders Across the Globe

"Felines" turned into a social peculiarity, breaking records and enamoring crowds around the world. The melodic's prosperity not just displayed the persevering through allure of Eliot's catlike characters yet in addition hardened the extraordinary force of writing when converted into the language of music, dance, and visual scene.

5. **Analysis and Understandings**

5.1 Abstract Legitimacy: Past A piece of cake

While "Down to earth Felines" is much of the time sorted as a work for youngsters, its scholarly legitimacy goes past simple a drop in the bucket.

Eliot's craftsmanship, phonetic smoothness, and capacity to catch the quintessence of feline ness raise the assortment to a level that rises above age hindrances. Pundits have praised Eliot for his innovative methodology and fun loving utilization of language.

5.2 Social Importance: Felines in the Social Embroidery

Eliot's "Pragmatic Felines" and the ensuing melodic transformation "Felines" have made a permanent imprint on mainstream society. The characters, stanzas, and songs have become imbued in the social woven artwork, impacting everything from images to spoofs and guaranteeing that Eliot's unconventional festival of cat erraticism perseveres in the shared mindset.

6. **Heritage: Paws and Verse Across Ages**

6.1 Getting through Notoriety: From Page to Execution

"Down to earth Felines" keeps on captivating perusers, everything being equal, demonstrating the getting through ubiquity of Eliot's capricious festival of cat characters. The outcome of the melodic transformation "Felines" further hardens the tradition of Eliot's reasonable felines, guaranteeing that these rough looking characters stay evergreen in the hearts of crowds.

6.2 Motivations and Tributes: Felines in Imaginative Works

Eliot's catlike sections have roused endless imaginative works, from representations and variations to reverences in writing and well known media. The characters of "Down to earth Felines" have become social standards, impacting specialists and authors across different kinds and mediums.

4.3 Edgar Allan Poe's Cat in "The Black Cat"

Edgar Allan Poe, a scholarly maestro of the grotesque, winds around a story of mental repulsiveness and extraordinary fear in "The Dark Feline." At the focal point of this chilling account is the baffling and premonition presence of a dark feline named Pluto. In looking at the story's perplexing layers, this investigation, crossing digs into the imagery, mental subtleties, and topical meaning of Poe's depiction of the catlike character. As Pluto turns into a mirror mirroring the tormented soul of the storyteller, the story discloses a perplexing dance between responsibility, franticness, and the ghostly secrets that hide inside the shadows.

1. **The Representative Presence of Pluto**

1.1 The Feline as a Representative Prime example

In writing, felines frequently convey representative weight, and on account of "The Dark Feline," Pluto arises as a powerful paradigm. Dark felines, saturated with legends and odd notion, frequently represent signs of incident or specialists of the powerful. Poe decisively takes advantage of this emblematic repository, utilizing Pluto to summon a feeling of premonition and to address further mental and otherworldly subjects.

1.2 Pluto: From Roman Folklore to Poe's Pen

The decision of the name Pluto isn't inconsistent. Gotten from the Roman divine force of the hidden world, Pluto, it foretells the plummet into obscurity that both the feline and the storyteller will embrace. This legendary association upgrades the story's inauspicious undercurrents and proposes an infinite arrangement of destiny, responsibility, and the powerful.

2. The Feline as an Impression of the Storyteller's Mind

2.1 Love Went to Detesting

At first, Pluto is a dearest ally to the storyteller. The feline's presence gives pleasure and solace to the family, filling in as a wellspring of fondness and kinship. In any case, the story takes a vile transform when the storyteller's plummet into liquor powered frenzy prompts a sickening demonstration of brutality: the gouging out of one of Pluto's eyes. This egregious demonstration denotes the start of a significant change in the storyteller's relationship with the feline.

2.2 A Reflection of Responsibility

As the storyteller capitulates to coerce and moral rot, so too does the idea of his relationship with Pluto. The feline, when a wellspring of comfort, turns into a living impression of the storyteller's tormented heart. The mutilation of Pluto's eye represents the storyteller's own ethical visual impairment, a self-caused wound that portends a drop into moral murkiness.

3. The Representative Versatility of Pluto

3.1 The Vanishing and Return

After the brutal distortion, Pluto's once-friendly disposition goes through a change. Rather than bringing out affection, the feline presently actuates dread and fear in the storyteller. However, the feline's emblematic versatility becomes evident as it won't leave the storyteller's side, typifying the inevitability of responsibility. Indeed, even as the storyteller endeavors to free himself of Pluto, the feline continues, tormenting him with a powerful persistence.

3.2 The Powerful Reverberation: The Ghostly Feline

Following Pluto's inopportune destruction, a subsequent feline enters the story — an almost indistinguishable dark feline looking similar to Pluto. This ghastly presence increases the story's otherworldly propensities, recommending a domain where the limits among life and passing, responsibility and retaliation, are liquid and unclear. The return of the feline turns into a ghastly reverberation of the storyteller's offenses, a persistent indication of the past.

4. The Feline as an Impetus for Frenzy

4.1 The Feline's Part in the Plummet

As responsibility consumes the storyteller, the ghostly feline turns into an impetus for the drop into franticness. Its appearances correspond with snapshots of significant mental unsettling influence for the storyteller, filling in as a harbinger of destruction. The common presence of the feline enhances the story's mental repulsiveness, obscuring the lines among the real world and the powerful.

4.2 A Baffling Shout: The Feline as Witness

In a climactic second, the unearthly feline coincidentally uncovers the storyteller's horrifying wrongdoing by emanating a piercing shout from inside the walls. This powerful indication highlights the feline's job as an unearthly observer to the unfurling misfortune. The catlike presence turns into a sign of responsibility and retaliation, rising above the domain of the normal and wandering into the strange and powerful.

5. The Mental Effect: Responsibility and Suspicion

5.1 The Mental Torture

Poe wonderfully makes a story that dives into the mental torture of the storyteller. Responsibility, similar to a tireless phantom, invades each feature of the storyteller's presence. The feline, with its emblematic reverberation and heavenly reverberations, turns into the epitome of the storyteller's interior torture, driving him to the edge of mental stability.

5.2 The Distrustfulness Heightens

As the story unfurls, the storyteller's distrustfulness heightens. He becomes persuaded that the phantom feline isn't simply a fantasy of his creative mind however a pernicious power with otherworldly organization. The obscuring of mental and powerful components uplifts the story strain, leaving perusers unsure about the real essence of the feline and the storyteller's plunge.

6. The Climactic Drop into Haziness

6.1 Homicide Generally Foul

In the peak of the story, the storyteller, headed to frenzy by the unearthly feline, carries out an unspeakable demonstration — killing his better half. This terrible development denotes the perfection of the storyteller's plummet into moral chasm. The extraordinary and mental strings meet in an embroidery of murkiness, leaving perusers on the cliff of loathsomeness and disclosure.

6.2 The Unpreventable Destiny

As the story rushes toward its decision, the storyteller regards himself as detained and anticipating the scaffold. The unpreventable results of his activities pose a potential threat, and the feline's otherworldly presence keeps on tormenting him. The feline's emblematic flexibility and otherworldly reverberations persevere, proposing that the results of culpability are inflexible, even past the limits of mortal presence.

7. Pluto's Getting through Inheritance

Edgar Allan Poe's "The Dark Feline" winds around a story of responsibility, franticness, and powerful fear, with the catlike character of Pluto filling in as a focal and puzzling figure. As an image, Pluto rises above the normal and turns into a mirror mirroring the storyteller's plunge into haziness. The feline's representative reverberation, combined with its extraordinary reverberations, adds layers of intricacy to the account, welcoming perusers to ponder the transaction of responsibility, profound quality, and the secrets that sneak inside the human mind.

"The Dark Feline" remains as a demonstration of Poe's capacity to intertwine mental frightfulness with the extraordinary, passing on a persevering through heritage that proceeds to enrapture and disrupt perusers.

Chapter 5

Cats In Art And Culture

Felines, those perplexing and effortless animals, have made a permanent imprint on human culture over the entire course of time. Their presence in workmanship and culture traverses landmasses and ages, rising above time and cultural changes. This exposition will dig into the multi-layered manners by which felines have been portrayed, worshipped, and coordinated into different types of imaginative articulation, as well as their effect on human social orders and convictions.

Authentic Points of view

The catlike interest traces all the way back to antiquated human advancements, where felines were frequently connected with divinities and supernatural characteristics. In antiquated Egypt, the feline was consecrated to the goddess Bastet, representing security, ripeness, and family life. Endless relics, from sculptures to wall canvases, deify the venerated cat structure, exhibiting the unmistakable job felines played in Egyptian workmanship.

In middle age Europe, felines took on an alternate job, frequently entwined with strange notions and legends. Dark felines, specifically, became related with black magic, prompting their oppression during the witch chases of the Medieval times. In any case, they likewise showed up in strict craftsmanship, as seen in enlightened compositions where felines were portrayed close by holy people or as allies to respectable figures.

Renaissance and Ornate Craftsmanship

As imaginative styles advanced, so did the portrayal of felines. The Renaissance denoted a re-visitation of old style impacts, and felines became well known subjects in both common and strict craftsmanship. Leonardo da Vinci, known for his sharp perception of nature, highlighted felines in a portion of his portrayals, catching their elegance and readiness.

The Florid time frame additionally enhanced the presence of felines in workmanship. Dutch painters, like Rembrandt and Vermeer, frequently remembered felines for their homegrown scenes, stressing the possibility of the feline as a sidekick in daily existence. The imagery joined to these catlike portrayals went from covertness and secret to homegrown warmth.

Japanese Workmanship and the Maneki-neko

In East Asia, especially in Japan, felines have held a novel and positive spot in social imagery. The Maneki-neko, or "enticing feline," is a well known Japanese doll accepted to carry best of luck and fortune to its proprietor. This notorious picture of a feline with a raised paw has risen above its social beginnings and turned into a worldwide image of thriving, frequently tracked down in shops and organizations all over the planet.

Japanese woodblock prints, or ukiyo-e, likewise habitually included felines. Utagawa Kuniyoshi, a prestigious ukiyo-e craftsman, made various prints displaying felines took part in different exercises, mirroring the creatures' perky and naughty nature.

Current and Contemporary Workmanship

The nineteenth and twentieth hundreds of years saw a different scope of creative developments, each offering an exceptional viewpoint on the portrayal of felines. The Post-Impressionist Henriette Ronner-Knip earned respect for her definite and loving canvases of felines, depicting them in homegrown settings with an extraordinary degree of authenticity and appeal.

In the domain of oddity, Salvador Dalí integrated cat symbolism into his fanciful and fantastical works. Felines, with their subtle and erratic way of behaving, fit consistently into Dalí's investigation of the psyche mind.

The contemporary workmanship scene keeps on being affected by felines, with specialists utilizing different mediums to investigate the perplexing connection among people and cats. Modern times has led to feline images and viral feline recordings, further setting the feline's place in mainstream society.

Writing and Felines

Past visual craftsmanship, felines have likewise assumed a huge part in writing. From antiquated tales like Aesop's "The Feline and the Mice" to T.S. Eliot's eccentric sonnets in "Old Possum's Book of Down to earth Felines," cats have been depicted as both clever and puzzling animals. Remarkably, Andrew Lloyd Webber adjusted Eliot's work into the famous melodic "Felines," which turned into a worldwide peculiarity.

Edgar Allan Poe's brief tale "The Dark Feline" digs into the more obscure parts of the catlike human relationship, investigating topics of culpability and franticness. The feline in this story turns into an image of the storyteller's inside disturbance, mirroring the diverse idea of the human-feline powerful in writing.

Felines in Mainstream society

In contemporary mainstream society, felines have become universal images, reflecting different aspects of human experience. From the Cheshire Feline in Lewis Carroll's "Alice's Experiences in Wonderland" to the puzzling Catwoman in comic books and movies, cat characters keep on enamoring crowds with their secret and charm.

In the realm of liveliness, felines have been highlighted in darling characters like Tom from "Tom and Jerry" and the famous Welcome Kitty, rising above social limits and becoming worldwide images of amusement and marketing.

Feline Love and Web Culture

Modern times has seen an extraordinary flood in feline related content. Images, viral recordings, and web-based entertainment accounts devoted to felines have become vital pieces of online culture. Cantankerous Feline, with her unendingly disappointed articulation, turned into a web sensation, generating a huge number of images and product.

Feline bistros, foundations where supporters can partake in the organization of occupant felines while tasting espresso, have acquired ubiquity in different areas of the planet. These spaces mix the social interest with felines with the contemporary craving for special and vivid encounters.

Imagery and Folklore

Felines have held emblematic importance in different legends and conviction frameworks. In Norse folklore, the goddess Freyja, related with adoration, excellence, and fruitfulness, rode in a chariot pulled by two enormous felines. The relationship among felines and womanliness, especially with goddesses, rises above societies and is clear in different legends around the world.

In Japanese old stories, the extraordinary animal Bakeneko is supposed to be a feline that has acquired powerful capacities, including the capacity to shape-shift into a human structure. These fantasies add to the getting through view of felines as baffling and otherworldly creatures.

5.1 Cats in Renaissance Art

The Renaissance, a time of social resurrection crossing from the fourteenth to the seventeenth hundred years in Europe, denoted an extraordinary period in human expression. Portrayed by a recharged interest in traditional learning, humanism, and logical request, the Renaissance significantly affected creative articulation. In the midst of the greatness of strict and fanciful topics, felines arose as captivating subjects in Renaissance workmanship, offering a brief look into the complex connection among people and these baffling cat mates.

Mainstream Movements in Renaissance Workmanship

The Renaissance saw a takeoff from the solely strict topics of the Medieval times, as specialists looked for motivation from traditional relic and the normal world.

This shift towards secularism considered a more extensive investigation of regular day to day existence and the consideration of common items and creatures in imaginative structures. Felines, with their elegant and baffling presence, tracked down a spot in the developing visual language of Renaissance workmanship.

Home life and Imagery

In the homegrown setting, felines became significant of the family and were many times remembered for scenes portraying day to day life. Works of art like Lorenzo

Lotto's "Representation of a Family" and Jan van Eyck's "Arnolfini Picture" unpretentiously consolidated felines as images of home life and friendship. These catlike figures, whether nestled into the hearth or energetically captivating with their human partners, added a component of warmth and commonality to the visual story.

Past simple portrayals of homegrown life, felines in Renaissance craftsmanship additionally conveyed representative importance. Their puzzling and frequently subtle nature made them charming subjects for craftsmen investigating further topics. Felines were saturated with characteristics that went from covertness and freedom to arousing quality and secret, adding layers of significance to the visual accounts in which they showed up.

Leonardo da Vinci and the Investigation of Nature

Leonardo da Vinci, the quintessential Renaissance polymath, made eminent commitments to the portrayal of felines in workmanship. Known for his careful investigations of the regular world, Leonardo's representations and drawings remembered nitty gritty perceptions of felines for different postures. These examinations displayed his sharp physical comprehension as well as caught the elegance and readiness that characterize the catlike structure.

In Leonardo's eminent painting, "Woman with an Ermine," a young lady holds an ermine, an image of virtue and home life. While not a homegrown feline, the ermine's incorporation mirrors the more extensive interest with creatures and the normal world during the Renaissance. Leonardo's interest stretched out to felines, and his point by point renderings stand as a demonstration of the period's expanding interest in logical perception.

Representations of Privileged

Felines were not restricted to the normal family; they additionally showed up in representations of the Renaissance gentry. Specialists depicted first class families with their pets, displaying a combination of riches, refinement, and an affection for creatures. These pictures, like Titian's "Representation of a Woman with a Lapdog," portrayed an amicable connection among people and their catlike mates, building up the possibility that felines were utilitarian as well as vital individuals from the family.

The presence of felines in privileged likeness likewise filled in as an image of refinement and complexity. Claiming and really focusing on fascinating or uncommon creatures, including felines, turned into a marker of status. The feline, with its detached disposition and smooth developments, added a component of style to these visual portrayals of the Renaissance tip top.

Feline as Moral story

Renaissance craftsmen regularly utilized moral story — an account gadget utilizing representative figures or activities — to convey more profound implications in their works. Felines, with their rich imagery, became vehicles for conveying complex thoughts and moral illustrations.

In the figurative composition "The Annunciation" via Carlo Crivelli, a feline is portrayed close to a mouse trap. This apparently unremarkable detail conveys representative weight, addressing the victory of good over evil. The feline, customarily connected with sly and covertness, fills in as a representation for carefulness and the capacity to defeat enticement.

Essentially, in Hans Memling's "Moral story of Virtuousness," a feline is shown playing with a spool of string, representing homegrown temperance and conjugal constancy. Here, the feline's perky connection with the string turns into a visual similitude for the sensitive equilibrium expected to keep up with virtue and dedication inside the homegrown circle.

Divine Cat: Felines in Strict Craftsmanship

While the Renaissance saw a shift towards common topics, felines actually tracked down their direction into strict craftsmanship, though in more nuanced ways. The feline, frequently connected with images of duality and secret, could be utilized to convey inconspicuous philosophical ideas.

In portrayals of the Nativity scene, craftsmen once in a while included felines as modest observers to the heavenly occasion. The juxtaposition of the heavenly and the customary, addressed by the presence of creatures like felines, served to refine and make engaging the sacrosanct accounts.

One prominent model is Leonardo da Vinci's "The Love of the Magi," where a feline is settled in the closer view, looking at the infant Christ. This consideration alludes to the comprehensiveness of the Nativity story, recommending that even animals as normal as felines give testimony regarding the exceptional situation transpiring within the sight of the Sacred Family.

Feline and Mouse: Imagery in Still Life

Renaissance still life works of art, a class that acquired conspicuousness during this period, frequently highlighted plans of ordinary items wealthy in emblematic importance. Felines and mice, habitually depicted together, became representative components in these organizations, addressing the timeless battle among great and malevolence, righteousness and bad habit.

In the work of art "Still Existence with Feline and Mouse" by Francesco Salviati, a feline is displayed in the demonstration of catching a mouse. The strain between the two animals adds a powerful component to the still life, while the representative ramifications address the fleetingness of life and the certainty of death.

Additionally, underway of craftsmen like Juan Sánchez Cotán and Jacopo de' Barbari, the consideration of felines and mice in still life structures added layers of significance, welcoming watchers to ponder the delicacy of presence and the recurrent idea of life and passing.

Feline Prints: Woodcuts and Inscriptions

The approach of printmaking during the Renaissance worked with the boundless spread of creative thoughts and pictures. Woodcuts and etchings highlighting felines became well known, considering a more extensive crowd to appreciate and draw in with cat symbolism.

Albrecht Dürer, an expert printmaker of the Renaissance, made a progression of woodcuts that remembered felines for different settings. His careful scrupulousness and capable delivering of the catlike structure raised the situation with felines in the domain of printmaking. These prints, frequently highlighting felines in energetic or pondering postures, became pursued masterpieces by their own doing.

5.2 The Cat in Japanese Ukiyo-e Prints

Japanese Ukiyo-e, the "photos of the drifting scene," addresses a type of workmanship that prospered during the Edo time frame (seventeenth to nineteenth hundreds of years) in Japan. Inside this energetic and different imaginative custom, felines arose as famous and enamoring subjects, typifying different social, representative, and tasteful aspects. This paper investigates the meaning of the feline in Japanese Ukiyo-e prints, following its development from straightforward portrayals of homegrown life to becoming notable images of fortune and appeal.

Homegrown Scenes and Regular day to day existence

In the beginning phases of Ukiyo-e, craftsmen like Hishikawa Moronobu and Suzuki Harunobu portrayed felines in scenes of day to day existence, frequently underlining their job as family friends. These prints, portrayed by sensitive lines and inconspicuous shading, exhibited the association among individuals and felines inside the cozy settings of homes and gardens.

Moronobu's prints, specifically, caught the conventional snapshots of homegrown life. Scenes of ladies playing with felines, taking care of them, or just noticing their fun loving shenanigans became famous subjects. The portrayal of felines in these prints served as a portrayal of day to day existence as well as an impression of the more extensive social shift towards valuing the basic delights and feel of the "drifting scene."

Utagawa Kuniyoshi: Felines in real life

Utagawa Kuniyoshi, a noticeable Ukiyo-e craftsman of the nineteenth 100 years, took the portrayal of felines higher than ever. Known for his dynamic and strong sytheses, Kuniyoshi's feline prints frequently highlighted these catlike animals in different conditions of activity — jumping, climbing, or took part in fun loving tricks.

One of Kuniyoshi's popular series, "One Hundred Writers, One Sonnet Each," incorporates a print highlighting a feline participated in a vivacious jump, catching a snapshot of dynamic energy. This takeoff from static depictions denoted a change in the creative treatment of felines, featuring their readiness and essentialness.

Kuniyoshi's prints additionally investigated the convergence of felines with incredible figures and legendary stories. Felines were depicted close by legends and champions, adding a component of caprice and appeal to customary stories. This mixing of the ordinary and the fantastical displayed Kuniyoshi's capacity to inject his prints with a feeling of story and energy.

Odd notion and Fables: Bakeneko and Nekomata

Japanese fables is rich with stories of powerful felines, and Ukiyo-e craftsmen jumped all over these legends to make spellbinding and creepy prints. The Bakeneko and Nekomata, legendary feline animals with powerful capacities, became well known subjects, mirroring a mix of strange notion and creative mind.

Bakeneko were accepted to be homegrown felines that procured enchanted powers in the wake of arriving at a particular age. In Ukiyo-e prints, Bakeneko were much of the time portrayed participating in human exercises, like wearing attire or performing moves. The combination of the common and the powerful in these prints mirrored the Japanese interest with the strange and the uncanny.

Nekomata, then again, were legendary feline animals with forked tails and the capacity to shape-shift into human structure. Ukiyo-e specialists depicted Nekomata in different appearances, accentuating their double nature as both customary felines and powerful creatures. These prints played on the pressure between the recognizable and the supernatural, spellbinding watchers with their portrayals of legendary cat substances.

Maneki-neko: The Coaxing Feline

The Maneki-neko, or "coaxing feline," is quite possibly of the most unmistakable and persevering through image in Japanese culture. Starting in the Edo period, the Maneki-neko is a charm accepted to carry favorable luck and thriving to its proprietor. Its unmistakable raised paw and alluring motion have made it a notorious portrayal of karma and appeal.

Ukiyo-e craftsmen embraced the prominence of the Maneki-neko, making prints that included this promising cat figure. These prints frequently depicted the Maneki-neko close by scenes of day to day existence, further solidifying its status as an image of homegrown flourishing. The feline, with its enticing paw, turned into a visual shorthand for welcoming favorable luck into the home.

The Maneki-neko's prevalence stretched out past Japan, and Ukiyo-e prints including this appealling feline tracked down a group of people across the globe. The picture of the enticing feline, with its wide-peered toward articulation and perky disposition, became inseparable from Japanese style and social character.

The Impact of Chinese Workmanship: Felines and Verse

Ukiyo-e specialists were affected by local Japanese practices as well as drew motivation from Chinese workmanship and writing. The Ming and Qing administrations in China highlighted feline themes in canvases and verse, rousing Japanese specialists to integrate comparative subjects into their own works.

Felines, representing polish and freedom, were frequently connected with lovely subjects in Chinese workmanship. Japanese Ukiyo-e craftsmen took on this theme, injecting their prints with a feeling of melodious magnificence. Felines were depicted in the midst of sprouting blossoms, pensively looking into the distance, encapsulating the amicable joining of nature and the homegrown domain.

One striking model is crafted by Utagawa Hiroshige, known for his scenes and idyllic structures. In his series "100 Popular Perspectives on Edo," Hiroshige included prints highlighting felines in the midst of cherry blooms and pre-winter leaves, making a graceful combination of the occasional and the representative.

Cherry Blooms and Felines: An Occasional Undertaking

Cherry blooms, or sakura, hold significant social importance in Japan, representing brevity and the excellence of fleeting minutes. Ukiyo-e specialists frequently matched cherry blooms with pictures of felines, making organizations that caught the short lived nature of life and the fragile appeal of the catlike structure.

In prints like Hiroshige's "Felines and Cherry Blooms," the juxtaposition of felines with blossoming cherry trees makes a feeling of idyllic despairing. The felines, with their pondering articulations, seem, by all accounts, to be as one with the evolving seasons, typifying the transient excellence of both nature and life.

This combination of occasional components with the appeal of felines represents the Ukiyo-e stylish, which commended the fleeting and the normal. The feline, in these prints, turns into a vessel for considering the recurrent idea of presence, welcoming watchers to ponder the progression of time and the getting through allure of the regular world.

5.3 Modern Cat Art and Its Impact on Pop Culture

In the contemporary workmanship scene, the universal cat has tracked down a spot as well as has arisen as a prevailing power, impacting patterns and catching the aggregate creative mind. Present day feline craftsmanship, with its underlying foundations in web culture, has risen above the limits of customary imaginative articulations, making a permanent imprint on mainstream society. This paper investigates the development of current feline workmanship, its different structures, and its significant effect on the manner in which we see and draw in with craftsmanship in the 21st 100 years.

The Web Feline Upheaval

The approach of the web carried with it a startling social peculiarity — the ascent of web feline culture. Images, viral recordings, and web-based entertainment stages turned into the material for displaying the appeal, wickedness, and peculiarities of our catlike companions. Felines, with their natural capacity to communicate a scope of feelings and their capricious way of behaving, immediately became web big names.

Grouchy Feline, maybe the web's most memorable cat sensation, caught the hearts of millions with her never-endingly disappointed articulation. Images highlighting Surly Feline spread like quickly, transforming her into a worldwide sensation and showing the web's ability to make moment superstars out of the most improbable subjects.

The prevalence of web feline culture made ready for a more extensive enthusiasm for felines in different types of media, making way for current feline craftsmanship to flourish. Images turned into a democratizing force, permitting anybody with a web association with add to the developing story of felines in mainstream society.

From Pixels to Paint: Web Feline Workmanship Rises above the Advanced Domain

As web feline culture picked up speed, specialists started to decipher the internet based interest with felines into conventional creative mediums. Advanced specialists, artists, and painters began making works that caught the quintessence of web feline culture while hoisting it to the domain of compelling artwork.

Pop surrealist craftsman Imprint Ryden, known for his unusual and fantastical manifestations, has regularly integrated felines into his intricate and fanciful creations.

His work of art "Rosie's Casual get-together" highlights a gathering of felines participating in a strange casual get-together, mixing components of imagination, sentimentality, and cat engage. Ryden's work exhibits the way that felines have become subjects as well as fundamental components in contemporary compelling artwork.

The progress from pixels to paint isn't restricted to laid out specialists; arising gifts on stages like Instagram and Etsy are additionally adding to the blossoming universe of feline craftsmanship. Free specialists and artists, motivated by the lively and expressive nature of felines, are making remarkable pieces that resound with crowds who look for both creative articulation and an association with the dearest cat sidekicks.

Hi Kitty: The Famous Impetus for Feline Promoting

The impact of felines on mainstream society stretches out past the domain of customary workmanship into promoting, with Hi Kitty remaining as the notable image of this cat roused business achievement. Made by Sanrio in 1974, Hi Kitty rises above ages and social limits. The person's moderate plan and general allure have transformed Hi Kitty into a worldwide peculiarity.

Hi Kitty's picture isn't bound to a particular item or market; it graces everything from writing material and dress to kitchenware and gadgets. The person's effect on mainstream society lies in its capacity to develop with the times while keeping an immortal appeal. The outcome of Hi Kitty shows the immense marketing potential that feline symbolism holds, affecting purchaser decisions and forming famous style.

In vogue: Felines on the Catwalk

The impact of current feline workmanship reaches out to the universe of style, where creators and brands have embraced cat themes, making a pattern that has caught the consideration of style lovers all over the planet. Feline themed dress, embellishments, and footwear have become staples in style assortments, mixing the appeal of felines with contemporary plan.

Top of the line creators like Karl Lagerfeld, known for his adoration for felines and his pet feline Choupette, have integrated cat themes into their assortments. Lagerfeld's plans frequently highlight feline molded frill, feline prints, and, surprisingly, whole assortments enlivened by the puzzling and lively nature of felines. The converging of high design with feline symbolism not just raises the situation with felines in the style world yet additionally mirrors a more extensive social shift toward embracing the capricious and the whimsical.

Past high style, feline themed clothing has turned into a standard pattern. Shirts embellished with adorable little cat delineations, sweaters highlighting complex feline examples, and adornments like feline formed gems have tracked down their direction into regular closets. This democratization of feline design addresses the far and wide allure and acknowledgment of felines as images of style and uniqueness.

Online Entertainment Feline Famous people: Another Sort of Impact

Online entertainment stages have led to another type of powerhouses — cat forces to be reckoned with. Felines with unmistakable characters and attractive appeal have amassed huge number of devotees, making them strong powerhouses by their own doing. Felines like Lil Pal, Nala Feline, and Smoothie the Feline have become easily recognized names, forming patterns and affecting social talk.

These virtual entertainment feline superstars give pleasure to their adherents as well as act as brand envoys. Organizations team up with these catlike powerhouses to advance items, taking advantage of the certifiable and positive affiliations that individuals have with felines. The impact of web-based entertainment feline superstars stretches out past web-based commitment, affecting purchaser conduct and building up the persevering through allure of felines in mainstream society.

Feline Bistros: A Social Peculiarity

Feline bistros have turned into a worldwide social peculiarity, joining the adoration for felines with the craving for extraordinary and vivid encounters. Starting in Japan, feline bistros have multiplied in significant urban communities around the world, furnishing supporters with the chance to partake in the organization of occupant felines while tasting espresso or tea.

The visual stylish of feline bistros frequently mirrors the impact of current feline craftsmanship. Wall paintings, works of art, and models of felines enhance the walls, making a climate that celebrates cat enchant. Feline bistros feature the combination of present day feline craftsmanship with certifiable encounters, permitting individuals to cooperate with felines in a fun loving and helpful climate.

These foundations not just add to the generally speaking social interest with felines yet in addition give a substantial and intelligent way for individuals to integrate the delight of felines into their day to day routines. The outcome of feline bistros highlights the significant effect of current feline craftsmanship on molding public spaces and social cooperations.

A Pawsitive Unrest

Present day feline craftsmanship has risen above the limits of customary creative articulations to turn into a pawsitive transformation that pervades each feature of mainstream society. From web images to artistic work magnum opuses, from design runways to feline bistros, the impact of felines has woven itself into the texture of contemporary society.

The democratization of feline culture, worked with by the web, has led to a more comprehensive and different portrayal of felines in workmanship and mainstream society.

Free craftsmen and arising gifts, alongside laid out names, keep on investigating the intricacies and fanciful notions of the human-feline relationship, making a rich embroidery of cat roused innovativeness.

As we explore the steadily advancing scene of present day feline workmanship, it becomes clear that felines have become something other than subjects; they are impetuses for satisfaction, motivation, and social movements. Whether in a gallery, on a design runway, or in the comfortable corner of a feline bistro, the effect of present day feline craftsmanship is a demonstration of the persevering and widespread allure of our catlike friends in the contemporary social outlook.

Chapter 6

Feline Celebrities

In the tremendous and different scene of mainstream society, a one of a kind and charming peculiarity has arisen — the ascent of cat famous people. Felines, with their enchanting shenanigans, strange appeal, and irrefutable moxy, have caught the hearts of millions around the world, rising above the conventional limits that once isolated human famous people from their catlike partners. This broad investigation digs into the interesting universe of cat superstars, looking at their beginnings, the variables that add to their fame, the effect they have on society, and the exceptional subculture that has arisen around these charming and shaggy stars.

The Development of Cat Notoriety

1.1 Verifiable Roots

The historical backdrop of cat big names can be followed back through the records of time, with felines possessing exceptional spots in the hearts and psyches of different societies. Old civic establishments, like the Egyptians, adored felines for their beauty and imagery, frequently depicting them close by divinities. The idea of felines as baffling and respected creatures established the groundwork for the social importance these animals would later achieve.

1.2 The Web Feline Transformation

The 21st century saw a groundbreaking time — the web feline transformation. With the approach of stages like YouTube, Instagram, and TikTok, felines became unforeseen yet stunningly famous web-based sensations. Images including felines, viral recordings displaying their shenanigans, and devoted virtual entertainment accounts transformed normal cats into web superstars. Any semblance of Testy Feline, Nyan Feline, and Console Feline accomplished uncommon popularity, demonstrating that the web was a strong balancer in the domain of big name.

Renowned Cat Appearances

2.1 Irritable Feline: The OG Web Sensation

Maybe the most famous figure in the web feline upheaval, Grouchy Feline (Tardar Sauce) rose to notoriety with her never-endingly displeased articulation. We investigate the story behind Cranky Feline, her effect on web culture, and the significant impact she had on forming the idea of cat famous people.

2.2 Lil Buddy: The Otherworldly Space Feline

Lil Buddy, with her novel appearance and charming character, turned into an image of flexibility and motivation. We dig into Lil Buddy's story, her excursion from being a salvage feline to a cherished VIP, and the beneficent undertakings she supported during her time at the center of attention.

2.3 Nala Feline: The Worldwide Powerhouse

Nala Feline, a staggering Siamese/Dark-striped cat blend, brags millions devotees via online entertainment stages. We investigate the ascent of Nala Feline, her job as a virtual entertainment powerhouse, and the effect she has had on molding patterns in feline driven content.

2.4 Maru: The Japanese YouTube Sensation

Hailing from Japan, Maru turned into a YouTube sensation with his cute and peculiar shenanigans, especially his affection for boxes. We look at the social meaning of Maru, his ubiquity in Japan and then some, and the persevering through allure of his enchanting and energetic disposition.

2.5 Smoothie the Feline: The Sovereign of Cushion

Smoothie, an English Longhair feline, acquired distinction for her lavish coat and superb disposition. We investigate the account of Smoothie, her ascent to notoriety on Instagram, and the special specialty she possesses in the realm of cat famous people.

The Feline Big name Way of life

3.1 Online Entertainment Fame

The job of online entertainment in catapulting felines to fame couldn't possibly be more significant. We dive into the mechanics of virtual entertainment superstar for felines, looking at the systems utilized by proprietors and administrators to organize drawing in happy, cooperate with crowds, and keep up with the appeal of their catlike stars.

3.2 Marketing and Brand Coordinated efforts

Cat superstars frequently broaden their impact past the advanced domain through marketing and brand joint efforts. We investigate how felines like Cantankerous Feline and Nala Feline have become brand representatives, loaning their pictures to a horde of items going from dress and accomplices to family things.

3.3 The Impact of Cat Superstars on Purchaser Conduct

Cat big names have the ability to shape buyer conduct and impact patterns. Through contextual investigations and market examinations, we inspect what the support of feline superstars means for the deals of feline related items and adds to the more extensive social interest with cats.

The Effect of Cat Big names on Society

4.1 Feline Reception and Salvage Backing

Numerous catlike superstars have utilized their acclaim to advocate for feline reception and salvage endeavors. We investigate the altruistic undertakings of felines like Lil Buddy and Nala Feline, revealing insight into how their big name status has been bridled to ultimately benefit their catlike partners.

4.2 The Restorative Job of Cat Famous people

The presence of cat famous people in the existences of their supporters goes past amusement — it has helpful ramifications. Research review and individual tributes feature the positive effect that watching and collaborating with feline famous people can have on mental prosperity.

4.3 Social Imagery and Portrayal

Cat superstars have become social images, addressing different characteristics and values. We dive into the social imagery related with felines and how cat famous people add to the portrayal of these representative implications in mainstream society.

The Feline VIP Subculture

5.1 Fan People group and Shows

The fanatics of cat VIPs structure dynamic networks, interfacing over their common love for these charming felines. We investigate the presence of fan networks, online discussions, and even shows where lovers assemble to celebrate and share their veneration for their number one cat famous people.

5.2 The Clouded Side: Abuse and Debates

The gigantic fame of cat VIPs has, now and again, prompted debates and moral worries. We analyze occurrences of double-dealing, contentions encompassing the treatment of VIP felines, and the obligation that accompanies dealing with the notoriety of these charming animals.

Cat Big names in Craftsmanship and Amusement

6.1 Cat Superstars as Dream

Cat superstars have motivated specialists and creatives across different mediums. We investigate how felines like Maru and Lil Buddy have become muses for craftsmen, bringing about a horde of works of art that commend their one of a kind characters.

6.2 Cat VIPs in Film and TV

The impact of cat VIPs stretches out to customary amusement mediums. We investigate how felines like Testy Feline and Nala Feline have shown up in films, TV programs, and ads, cementing their status as hybrid stars.

6.3 Cat Famous people in Writing and Marketing

The appeal of cat famous people has pervaded writing and marketing. From kids' books highlighting renowned felines to stock bearing their pictures, we look at the different manners by which these catlike superstars have become vital to different types of creative articulation.

The Eventual fate of Cat VIP

7.1 Patterns and Advancements

As innovation advances and cultural inclinations shift, the universe of cat superstars is probably going to see recent fads and developments. We investigate likely future turns of events, from progressions in computer generated simulation encounters to imaginative showcasing techniques that profit by the persevering through prevalence of cat big names.

7.2 Moral Contemplations and Mindful Possession

The rising perceivability of cat famous people raises significant moral contemplations in regards to mindful proprietorship, creature government assistance, and the likely traps of acclaim. We dig into the developing scene of moral contemplations encompassing the administration of cat big names and the obligations that accompany their popularity.

6.1 The Cat From Outer Space: A Look at Cats in Sci-Fi

Felines have long held a strange and extraordinary charm, making them a characteristic fit for the speculative domains of sci-fi. From old legends to contemporary film, cats have frequently been related with enchantment, otherworldliness, and a dash of the extraterrestrial. This investigation digs into the depiction of felines in sci-fi, analyzing how these cryptic animals cross the universe, collaborate with modern advancements, and become necessary to the accounts of powerful undertakings.

Felines as Inestimable Sidekicks

1.1 Antiquated Roots: Felines in Folklore

The relationship among felines and the powerful originates before present day sci-fi, with old developments integrating cats into their legends and strict convictions. Felines were frequently connected to divinities, representing both elegance and secret. In investigating these old roots, we gain experiences into the persevering through interest with felines as creatures that rise above the natural domain.

1.2 From Old Egypt to the Stars

The love for felines went on in Antiquated Egypt, where they were appreciated mates as well as worshipped as images of security and favorable luck. This heritage laid the preparation for the depiction of felines in sci-fi, where they flawlessly progress from natural allies to astronomical travelers.

Cat Intergalactic Travelers

2.1 The Feline From Space (1978)

The film "The Feline From Space" fills in as an early instance of mixing cat fascinate with intergalactic experiences. We investigate how this Disney exemplary made ready for the depiction of felines as extraterrestrial creatures, investigating subjects of kinship, innovation, and the strange powers that felines could have.

2.2 Catnapped in Space: On board Spaceships and Then some

Various sci-fi works have included felines as startling stowaways on board spaceships. Whether it's Jonesy in "Outsider" (1979) or Spot in "Star Journey: The Future" (1987-1994), we investigate the repetitive theme of felines exploring the universe, adding a bit of warmth and predictability to the cool endlessness of room.

Felines and Outsider Experiences

3.1 The Egyptian Mau in "Men dressed in Dark" (1997)

In "Men dressed in Dark," the Egyptian Mau, a genuine variety known for its unmistakable spots, assumes a critical part in the cooperations among people and extraterrestrial creatures. We analyze how the film utilizes the persona of the Egyptian Mau to represent the association among Earth and space.

3.2 Schrödinger's Feline: Quantum Felines and Substitute Real factors

In the domain of hypothetical material science, Schrödinger's feline has turned into a psychological test investigating the idea of quantum superposition. We dig into how this logical idea has tracked down its direction into sci-fi, with felines becoming similitudes for the vulnerabilities and potential outcomes of substitute real factors.

Feline astrophes and Whole-world destroying Cats

4.1 The Feline as Harbinger of Destruction

In some sci-fi stories, felines take on whole-world destroying jobs, filling in as harbingers of destruction or otherworldly creatures attached to disastrous occasions. We investigate stories where felines become influencers, proclaiming the finish of universes or directing heroes through tragic scenes.

4.2 "Nine Lives" (2016): Cat Interminability and the Apocalypse

The film "Nine Lives" acquaints a comedic curve with the prophetically calamitous feline story, where a business magnate winds up caught in the body of a feline. We break down how the film plays with the saying of cat interminability and its suggestions for both the person and the world at large.

Cat Cyborgs and Techno-Felines

5.1 The Bionic Feline: Cyborg Cats in Science fiction

As innovation progresses, sci-fi investigates the combination of felines and robotics. We look at stories where felines become bionic creatures, exploring modern scenes with upgraded capacities and filling in as an impression of the developing connection among mankind and innovation.

5.2 "A.I. Love You" (1994): Virtual Felines and Advanced Domains

The manga "A.I. Love You" presents the idea of virtual felines, obscuring the lines between the physical and advanced universes. We investigate how the story dives into the special friendship among people and virtual cat elements, bringing up issues about the idea of the real world and association.

Felines in Science fiction Writing

6.1 Philip K. Dick's "Do Androids Long for Electric Sheep?" (1968)

Philip K. Dick's persuasive novel investigates the topic of compassion through the personality of Deckard's electric sheep. We examine the imagery of creatures, including felines, in reality as we know it where manufactured creatures question the limits among natural and counterfeit life.

6.2 Neal Stephenson's "Snow Crash" (1992)

In "Snow Crash," a cyberpunk novel, the person Y.T. experiences a feline that ends up being a bioengineered monitor creature. We dig into the job of felines in this tragic future, where biotechnology obscures the lines between the normal and the made.

The Similitude of the Feline

7.1 Freedom and Secret: Felines as Similitudes

Felines in sci-fi frequently act as analogies for freedom, secret, and the untamed parts of the universe. We investigate how the figurative characteristics of felines upgrade the account profundity of sci-fi works, furnishing layers of implying that resound with both the characters and the crowd.

7.2 Feline Individuals: Shapeshifting and Transformation

In some sci-fi stories, felines exemplify the idea of shapeshifting and transformation. We dissect stories where cat characters go through changes, investigating subjects of personality, versatility, and the ease of presence.

The Fate of Felines in Science fiction

8.1 Arising Patterns and Topics

As sci-fi keeps on developing, we expect to arise patterns and subjects connected with felines. Whether it's the investigation of cutting edge hereditary designing, computer generated simulation encounters based on cat friendship, or imaginative stories that challenge conventional ideas of the real world, what's to come holds energizing opportunities for felines in the speculative domain.

8.2 Moral Contemplations: The Treatment of Science fiction Felines

With the rising presence of felines in sci-fi, moral contemplations emerge in regards to their treatment inside these stories. We dive into the obligations of makers and narrators in creating accounts that regard the government assistance and organization of cat characters.

6.2 Famous Internet Cats: Grumpy Cat, Lil Bub, and More

The web, a huge and apparently endless space, has birthed its own heavenly body of stars, and among them, none sparkle as splendidly as well known web felines. These catlike big names, with their particular characters and certain appeal, have enthralled crowds around the world, rising above the computerized domain to become notorious figures in mainstream society. This investigation dives into the lives and traditions of probably the most well known web felines, including Cranky Feline, Lil Pal, and a couple of other people who have left a permanent pawprint on the internet.

Cranky Feline - The OG Web Sensation

1.1 Ascent to Fame

Cranky Feline, whose genuine name was Tardar Sauce, arose as the OG (Unique Crab) web sensation. With a never-endingly displeased articulation brought about by cat dwarfism, Irritable Feline's famous look turned into an unexpected phenomenon when a photograph of her was posted on Reddit in 2012. We dive into the conditions that prompted Grouchy Feline's ascent to fame and how her special appearance caught the hearts of millions.

1.2 Images, Product, and Media Appearances

Irritable Feline's ubiquity soar as her picture turned into the dream for endless images and viral substance. The surly disposition, matched with diverting inscriptions, produced a plenty of web humor. Grouchy Feline rose above the advanced domain, showing up on stock, in advertisements, and in any event, featuring in her own Lifetime film, "Surly Feline's Most obviously terrible Christmas Ever." We investigate the effect of Cantankerous Feline's web notoriety on different parts of mainstream society.

Lil Buddy - The Supernatural Space Feline

2.1 An Exceptional Appearance

Lil Buddy, brought into the world as a half-pint with a few hereditary peculiarities, had a particular appearance described by her little size, interesting facial elements, and an unendingly hanging tongue. We dig into how Lil Pal's remarkable actual attributes added to her appeal and set before her the way to web fame.

2.2 Lil Pal's Internet based Realm

Lil Pal's web presence reached out past simple reverence; it turned into a stage for inspiration and generosity. With a devoted following, Lil Pal's proprietor, Mike Bridavsky, used her distinction to bring issues to light and assets for creature good cause. We investigate how Lil Pal's web-based realm turned into a power for good, displaying the positive effect that web feline superstars can have on worthy missions.

Nyan Feline - A Mainstream society Peculiarity

3.1 The Introduction of Nyan Feline

Nyan Feline, a pixelated cat with a Pop-Tart body leaving a rainbow limp along, turned into a web image that rose above language and social obstructions. We investigate the beginnings of Nyan Feline, following its creation back to a YouTube video in 2011, and how it immediately turned into a worldwide mainstream society peculiarity.

3.2 From Image to Product

Nyan Feline's irresistible and snappy nature prompted a multiplication of product, going from dress and accomplices to toys and computer games. We analyze how Nyan Feline's excursion from image to stock epitomizes the groundbreaking force of web images in forming mainstream society and purchaser patterns.

Console Feline - The Maestro of Images

4.1 The Image that Resounded

Console Feline, with its snappy tune and film of a feline apparently playing an electronic console, became one of the early web images that resounded with a worldwide crowd. We investigate the beginnings of Console Feline, its development into an image, and how it turned into an image of web humor.

4.2 Persevering through Heritage

In spite of the death of the first Console Feline, the image's heritage lives on as an image of wistfulness and web culture. We talk about how Console Feline made ready for other feline based web peculiarities and made a persevering through imprint on the image scene.

Pudge - The Sovereign of Objection

5.1 A Face That Says Everything

Pudge, a Scottish Overlay feline with a novel appearance and a particular objecting articulation, cut out her own specialty in the realm of web feline big names. We dig into how Pudge's looks and particular markings added to her status as the "Sovereign of Dissatisfaction."

5.2 The Convergence of Backtalk and Style

Pudge not just turned into a web sensation for her opposing looks yet additionally acquired prevalence for her chic photoshoots. We investigate how Pudge's convergence of backtalk and style caught the consideration of a devoted fan base and raised her status as an in vogue cat.

Smoothie the Feline - The Sovereign of Cushion

6.1 Lavish Locks and Striking Eyes

Smoothie, an English Longhair feline, separated herself with her lavish fur and enrapturing blue-green eyes. We investigate how Smoothie's actual traits, joined with her superb disposition, transformed her into the "Sovereign of Cushion" and procured her a noticeable spot among popular web felines.

6.2 Web-based Entertainment Fame

Smoothie's ascent to fame was powered by her enrapturing photographs shared on Instagram. We break down how online entertainment stages, especially Instagram, have become amazing assets for catapulting felines like Smoothie to worldwide notoriety and developing a worldwide local area of cat lovers.

Feline Big names as Forces to be reckoned with

7.1 Virtual Entertainment Effect

The impact of these popular web felines stretches out past simple amusement; they have become powerhouses with the capacity to shape drifts and draw in with enormous crowds. We investigate the effect of these catlike powerhouses via virtual entertainment, dissecting how they influence their characters to associate with fans and advance different items.

7.2 Brand Joint efforts

Popular web felines frequently work together with brands, making a cooperative energy between cat appeal and business offer. We dig into how coordinated efforts with brands add to the monetary progress of feline powerhouses and the developing scene of force to be reckoned with advertising inside the domain of web felines.

The Clouded Side of Web Feline Notoriety

8.1 Abuse and Moral Contemplations

The notoriety of web felines has a clouded side, with occasions of double-dealing, exploitative treatment, and contentions encompassing their prosperity. We address the moral contemplations related with dealing with the notoriety of renowned web felines and the obligation that falls on proprietors, chiefs, and the more extensive internet based local area.

8.2 Lamenting the Deficiency of Web Feline Big names

The death of well known web felines, like Cranky Feline and Lil Pal, evokes a special type of internet grieving. We analyze how the web local area processes and laments the deficiency of cherished cat superstars, thinking about the significant effect these felines have had on the existences of their fans.

6.3Cats on the Silver Screen: From "Breakfast at Tiffany's" to "Puss in Boots"

Felines have been a staple in film, gracing the cinema with their appeal, secret, and once in a while, their devilish shenanigans. From exemplary movies to energized experiences, felines play played different parts, becoming famous characters in the realm of film.

This investigation takes us on a realistic excursion, following the depiction of felines in critical movies, from the immortal tastefulness of "Breakfast at Tiffany's" to the vivified eccentricity of "Puss in Boots."

"Breakfast at Tiffany's" (1961) - Holly Golightly's Catlike Sidekick

1.1 The Notorious Feline: Orangey as Feline

"Breakfast at Tiffany's," coordinated by Blake Edwards and in view of Truman Overcoat's novella, includes a notorious cat character just known as Feline. Played by the gifted feline entertainer Orangey, Feline turns into the quiet ally to Holly Golightly, depicted by Audrey Hepburn. We investigate how Feline represents something beyond a pet, filling in as a similitude for Holly's yearning for a feeling of having a place and solidness.

1.2 Feline as an Image of Freedom

In "Breakfast at Tiffany's," Feline exemplifies the soul of autonomy, reflecting Holly's craving for opportunity and self-articulation. We dig into the imagery of Feline as a quiet friend, underscoring the extraordinary connection among people and their catlike colleagues and how this bond is depicted as a fundamental component in the film.

"The Aristocats" (1970) - Vivified Style and Melodic Party

2.1 Disney's Catlike Privileged

Walt Disney's "The Aristocats" acquaints crowds with a cast of vivified cat characters exploring the roads of Paris. Duchess, Thomas O'Malley, and their little cats become the overwhelming focus in this endearing story of experience, music, and cat fellowship. We investigate how Disney's vivified include brings a dash of class and melodic excess to the universe of feline driven narrating.

2.2 Subjects of Family and Versatility

"The Aristocats" winds around a story that underlines the significance of family bonds and versatility. Through the preliminaries and experiences looked by the catlike characters, the film catches the substance of felines as both autonomous and profoundly associated animals. We break down how Disney's depiction of cat effort-lessness and appeal adds to the ageless allure of "The Aristocats."

"Feline Individuals" (1942) - A Story of Loathsomeness and Change

3.1 Exemplary Ghastliness with Cat Curve

Coordinated by Jacques Tourneur, "Feline Individuals" is an exemplary blood and gore movie that investigates the extraordinary association among people and cats. We dig into how the film presents the idea of a revile that changes its characters into lethal pumas, mixing frightfulness with an investigation of human cravings and basic impulses.

3.2 Allegorical Imagery

"Feline Individuals" utilizes the representation of cat change to dig into mental and sexual subjects. The film investigates the duality of human instinct, drawing matches between the baffling and charming characteristics of felines and the secret profundi-ties inside people. We break down the mental suggestions and figurative imagery that make "Feline Individuals" a special section in cat roused film.

"The Feline Returns" (2002) - An Enlivened Japanese Experience

4.1 Studio Ghibli's Catlike Dream

"The Feline Returns," a Studio Ghibli movie coordinated by Hiroyuki Morita, takes crowds on a capricious excursion into an otherworldly feline realm. The film investigates the results of a normal young lady being cleared into a universe of talking felines and experience. We inspect how Studio Ghibli mixes its remarkable kind of sorcery and creative mind into the story, making a spellbinding story for crowds, everything being equal.

4.2 Subjects of Strengthening and Self-Revelation

"The Feline Returns" dives into subjects of strengthening and self-revelation as the hero, Haru, explores the difficulties of the feline realm. Through her collaborations with cat characters, the film commends finding one's inward strength and personality. We investigate the groundbreaking excursion of Haru and the captivating universe of cat enchantment in Studio Ghibli's narrating.

"Puss in Boots" (2011) - Brave Experiences in the Shrek Universe

5.1 Puss in Boots: The Catlike Swashbuckler

"Puss in Boots," a side project from the Shrek establishment, puts the focus on the trying and charming cat character, voiced by Antonio Banderas. We investigate how Puss in Boots' personality, at first presented as a supporting figure in Shrek, turned into a fan number one, prompting a film that exhibits his daring experiences and gives a funny interpretation of exemplary fantasy sayings.

5.2 The Unique Team: Puss and Humpty Dumpty

"Puss in Boots" presents a one of a kind dynamic among Puss and his previous com-panion turned foe, Humpty Dumpty. The film investigates subjects of companion-ship, selling out, and reclamation, winding around a story that goes past the ordinary fantasy. We dive into the person elements and the humor that characterize the film's way to deal with cat focused narrating.

"A Road Feline Named Sway" (2016) - Genuine Motivations

6.1 An Inspiring Genuine Story

"A Road Feline Named Sway" depends on the genuine story of James Bowen and his took on feline, Weave. We investigate how the film brings a genuine story of recovery and friendship to the screen, underscoring the effect of cat sidekicks on living souls.

6.2 The Mending Force of Felines

The film depicts the mending force of the connection among James and Weave, displaying how the presence of a feline can carry positive change to the existences of people confronting difficulties. We dissect how "A Road Feline Named Bounce" remains as a demonstration of the extraordinary job felines play in giving solace, friendship, and a feeling of direction.

Cat Stars in Film: Getting through Allure and Social Importance

7.1 Felines as Representative Characters

Across these movies, felines act as representative characters, addressing a heap of subjects from freedom and change to friendship and strengthening. We investigate the getting through allure of felines in film and the social meaning of their depiction in narrating.

7.2 The Diverse Idea of Felines

The depiction of felines on the cinema mirrors the diverse idea of these animals — strange, free, enchanting, and some of the time, mysterious. We analyze how producers use felines as adaptable characters, improving accounts with their interesting characteristics and adding layers of significance to artistic narrating.

Chapter 7

Cats in Warfare

Felines, with their autonomous and dexterous nature, have ended up laced in the archives of fighting over the entire course of time. From old civilizations to current contentions, these perplexing cats play played different parts, filling both functional and emblematic needs on the combat zone. This exhaustive investigation dives into the complex connection among felines and fighting, inspecting their jobs as friends, bug regulators, and even images of karma and odd notion.

Felines in Old Human advancements

1.1 Egypt: Watchmen of the Pharaohs

In old Egypt, felines held a consecrated status and were related with the goddess Bastet, the defender of home and family. Felines were loved as well as effectively utilized in defending storehouses and food supplies from rodents. We investigate the social meaning of felines in old Egyptian culture and their parts in safeguarding significant assets.

1.2 Rome and Greece: Cat Mates in Armies

In old Rome and Greece, felines filled in as allies to warriors and mariners, giving both reasonable nuisance control and basic encouragement. We dig into the authentic records and relics that portray the presence of felines in the tactical societies of these civilizations, revealing insight into the bond fashioned among people and cats on the bleeding edges.

Felines in Middle age Europe

2.1 Shipboard Mates and Irritation Regulators

During the Medieval times, felines became apparatuses on European boats, helping mariners in keeping rat populaces under control. The sea custom of having boat's felines endured as the centuries progressed, with these catlike team individuals giving useful advantages and a bit of solace during long and dangerous excursions.

2.2 Notions and Imagery

Felines likewise became images of odd notion in archaic Europe, with convictions encompassing their alleged supernatural properties. Dark felines, specifically, were related with black magic, prompting both veneration and dread. We investigate what these notions meant for the treatment of felines and their jobs in archaic military settings.

Felines in Wartime Naval forces

3.1 The Time of Sail: Felines as Rodent Catchers

As sea exchange extended during the Period of Sail, felines assumed a significant part in battling the wild rat populaces on ships. Their presence guaranteed a better living climate as well as added to the protection of food supplies. We analyze first-hand records and verifiable records specifying the irreplaceable commitments of boat's felines in maritime settings.

3.2 Maritime Mascots and Buddies

Past their utilitarian jobs, felines on board sends frequently became cherished mascots and allies to mariners. These catlike shipmates gave a feeling of fellowship and solace in the brutal and eccentric states of maritime life. We investigate the accounts of eminent maritime felines and their getting through influence on sea customs.

Felines in Fighting: An Image of Karma

4.1 Fortunate Felines in Asian Societies

In different Asian societies, felines are viewed as images of best of luck and flourishing. The "Maneki-neko" or coaxing feline, frequently portrayed with a raised paw, is a typical charm accepted to bring favorable luck. We dive into how these images tracked down their direction into military settings, filling in as defensive charms and signs of progress.

4.2 Felines in Military Fables

Felines have likewise tracked down a spot in military legends, with accounts of their mysterious capacities and defensive presence flowing among troopers. From warding off abhorrent spirits to anticipating approaching risks, felines have become legendary figures interwoven with the fables of military customs. We investigate the social meaning of felines as images of karma and guardianship.

Felines as Spies and Couriers

5.1 Surveillance and Secretive Tasks

In a few verifiable occasions, felines were used for reconnaissance purposes. Their secrecy and capacity to explore new territory made them powerful specialists in undercover tasks. We dig into models where felines were prepared or just took advantage of for their regular impulses in get-together knowledge during seasons of war.

5.2 Courier Felines: Paws in Correspondence

Felines, known for their homing impulses, were utilized as couriers in specific military circumstances. Their capacity to cross troublesome territory and convey messages across adversary lines demonstrated significant in times when more customary method for correspondence were unrealistic or hazardous. We investigate how felines filled in as a novel and frequently ignored part of military correspondence.

War Felines in the Advanced Time

6.1 Felines in Close quarters conflict

During The Second Great War, felines ended up in the midst of the bleak real factors of close quarters conflict. Fighters down and dirty kept felines as partners, giving comfort and brotherhood amidst the revulsions of war. We look at the individual records and photos that feature the presence of felines on the bleeding edges.

6.2 Felines in The Second Great War

The Second Great War saw the continuation of felines as the two colleagues and nuisance regulators in military settings. Their jobs reached out past reasonable items, offering everyday encouragement to troopers in testing and upsetting conditions. We investigate the wartime encounters of felines and the connections manufactured with servicemen and ladies.

The Contemporary Job of Felines in the Military

7.1 Military Working Felines

In present day times, felines keep on assuming parts in military conditions, especially in controlling irritations around army bases. We investigate how military functioning felines add to keeping up with cleanliness and protecting assets, repeating their verifiable jobs as irritation regulators in the midst of contention.

7.2 Treatment Felines and Daily encouragement

Felines have additionally become important supporters of treatment and daily reassurance programs inside the military. Perceiving the positive effect of creatures on psychological wellness, treatment felines are utilized to give solace and friendship to support individuals managing the burdens of military life.

The Morals of Including Felines in Fighting

8.1 Creature Government assistance and Moral Contemplations

The utilization of felines in fighting raises moral contemplations connected with creature government assistance. We dig into the discussions encompassing the contribution of creatures, including felines, in military exercises, investigating the points of view on the moral treatment and utilization of creatures in struggle zones.

8.2 Promotion and Guidelines

Basic entitlements activists and associations advocate for the others conscious treatment of creatures in all unique circumstances, including fighting. We look at the administrative systems and rules that address the moral contemplations of including felines in military exercises and the continuous endeavors to guarantee the prosperity of creatures in such conditions.

Felines in Mainstream society and Media

9.1 Felines in War Writing

The presence of felines in fighting has been portrayed in different types of writing. From war books to verse, writers have investigated the mind boggling connections among fighters and their catlike sidekicks. We dissect how war writing plays depicted the parts of felines and the close to home bonds shaped in the midst of contention.

9.2 Felines in War Movies and Craftsmanship

War films and creative portrayals frequently remember felines as emblematic or strict components for their accounts. From depicting the difficulties of battle to representing trust and flexibility, we investigate how felines have been portrayed in clear line of sight and realistic expressions as impressions of the human involvement with seasons of contention.

7.1 Ship Cats: Cats at Sea Throughout History

The immense field of the untamed ocean, with its erratic waves and huge skylines, has been a space shared by people and a one of a kind gathering of mates: transport felines. These catlike mariners have been a necessary piece of sea history, going with sailors on their journeys across seas and filling in as both viable irritation regulators and appreciated friends. This thorough investigation digs into the rich woven artwork of boat felines since forever ago, from their jobs on antiquated vessels to their getting through presence on current boats.

The Beginnings of Boat Felines

1.1 Felines in Old Sea Societies

The connection among felines and mariners has old roots, tracing all the way back to nautical societies like the Phoenicians and Greeks. Felines were esteemed for their capacity to control rodent and mouse populaces on board delivers, protecting food supplies and guaranteeing the security of vessels. We investigate the early occurrences of felines as oceanic sidekicks and the viable jobs they played in old nautical social orders.

1.2 Egyptian Nautical Felines: Watchmen of Exchange

In antiquated Egypt, felines were adored and considered consecrated creatures related with security. As the Egyptians participated in sea exchange, felines became fundamental buddies on ships, offering both viable help and filling in as images of favorable luck. We dig into the verifiable records and antiquities that shed light on the presence of felines in old Egyptian marine.

Felines on the High Oceans in the Time of Investigation

2.1 The Renaissance and the Time of Investigation

As European powers set out on aggressive journeys of investigation during the Renaissance, felines wound up on board the vessels of popular pilgrims like Christopher Columbus and Ferdinand Magellan. Their jobs extended past vermin control, giving solace and friendship to mariners confronting the difficulties of strange waters. We investigate the advancing jobs of felines during this period of oceanic investigation.

2.2 Privateers and Privateers: Felines in Sea Mischief

On the uncivilized oceans of the Brilliant Period of Robbery, felines became installations on privateer and privateer ships. With their sharp faculties and capacity to adjust to deliver life, felines assumed a double part as colleagues and caution frameworks, making groups aware of the presence of undesirable guests. We dig into the vivid accounts of privateer felines and their place in oceanic rebels' displays.

Boat's Felines in the Time of Sail

3.1 The Nautical Custom: Felines On board Maritime Vessels

During the Time of Sail, maritime vessels became home to various boat felines, procuring their keep as master rodent catchers. The Imperial Naval force, specifically, embraced the practice of having felines on board its boats, perceiving their priceless commitments to keeping a sound and sterile boat climate. We investigate how felines turned into a basic piece of the nautical custom during this period.

3.2 Shipboard Notions: Felines as Images of Best of Luck

Mariners, known for their notions, believed felines to be images of best of luck. Dark felines, specifically, were accepted to bring fortune and shield ships from hurt. We unwind the notions and oceanic fables encompassing boat felines, looking at the ceremonies and convictions that mariners maintained to guarantee safe journeys.

Well known Boat Felines in History

4.1 Trim: Matthew Flinders' Catlike Pioneer

Trim, the boat feline of English pilot Matthew Flinders, went with him on his notable circumnavigation of Australia in the mid nineteenth 100 years. Trim's endeavors and undertakings on board the HMS Examiner became amazing, and his story embodies the nearby connection among mariners and their catlike sidekicks.

4.2 Resilient Sam: The Feline with Different Lives

Resilient Sam, otherwise called Oscar or Oskar, accomplished a surprising accomplishment by enduring the sinking of three distinct boats during The Second Great War — the Bismarck, the HMS Cossack, and the HMS Ark Regal. We investigate the fantastic story of this tough boat feline and his excursion through a portion of the conflict's most critical maritime commitment.

Transport Felines in the Advanced Time

5.1 Maritime Felines in Universal Conflicts

Felines kept on assuming parts in maritime activities during the twentieth 100 years, with their presence reported on warships during both The Second Great War and The Second Great War. Their commitments went from bother control to lifting the general mood among mariners. We look at the authentic records and tales that feature the persevering through job of boat felines in current maritime history.

5.2 The Decay and Resurgence of Boat Felines

As innovation progressed and maritime practices advanced, the conventional job of boat felines lessened. Nonetheless, lately, there has been a resurgence of interest in having felines on board vessels, for viable reasons as well as for the positive effect they can have on group prosperity. We investigate contemporary cases of boat felines and the explanations for their re-visitation of sea conditions.

Felines and Nautical Old stories

6.1 Felines as Marine Images

Nautical fables is packed with stories and odd notions including felines. From their alleged capacity to foresee tempests to their relationship with favorable luck, felines have become getting through images of the oceanic world. We disentangle the legends and stories that have added to the persona encompassing boat felines in nautical fables.

6.2 The Feline O' Nine Tails: A Catlike Whipping Legend

The expression "feline o' nine tails" is frequently connected with a kind of whip utilized in maritime discipline. As opposed to mainstream thinking, this has no immediate association with cat discipline. We investigate the starting points of the term and dissipate the legend encompassing the feline o' nine tails.

Felines as Nautical Dream and Motivation

7.1 Felines in Sea Writing

The presence of felines on ships has motivated various works of writing, from ocean shanties and sonnets to books and brief tales. We dive into the scholarly tradition of boat felines, investigating how these catlike colleagues have been woven into the texture of sea narrating.

7.2 Boat Felines in Workmanship and Photography

The charm of boat felines reaches out to the domain of visual expressions, with portrayals of felines on board vessels showing up in artistic creations, delineations, and photos. We investigate how craftsmen and picture takers have caught the pith of boat felines, deifying their parts in sea history.

The Tradition of Boat Felines

8.1 Respecting Boat Felines: Remembrances and Recognitions

In acknowledgment of the persevering through commitments of boat felines, different remembrances and recognitions have been laid out to respect their heritage. From sculptures to plaques, these recognitions commend the novel connection among mariners and their catlike colleagues. We investigate the manners by which boat felines are recollected and celebrated all over the planet.

8.2 Boat Felines in Mainstream society

The tradition of boat felines lives on in mainstream society, with references showing up in movies, writing, and, surprisingly, virtual entertainment. From fictionalized records to narratives, we investigate how boat felines keep on catching the creative mind of crowds and keep up with their status as charming sea figures.

The Moral Treatment of Boat Felines

9.1 Government assistance Contemplations

While transport felines generally served down to earth jobs on board vessels, the moral treatment of creatures has turned into a central concern. We look at the moral contemplations encompassing the presence of felines on ships, resolving issues, for example, their prosperity, veterinary consideration, and the obligations of the people who decide to have felines as a component of the group.

9.2 Present day Rules and Guidelines

In the contemporary oceanic industry, rules and guidelines oversee the presence of creatures on ships. We investigate the current systems that expect to guarantee the altruistic treatment of boat felines, accentuating the significance of offsetting custom with moral obligation.

7.2 Cats in Ancient and Medieval Warfare

Felines, worshipped for their baffling and free nature, assumed critical parts in the woven artwork of old and archaic fighting. From the great multitudes of old civilizations to the wild clashes of the middle age time frame, felines went with troopers, offering pragmatic answers for bug control, filling in as sidekicks, and, surprisingly, expecting representative importance. This investigation digs into the multi-layered connections among felines and fighting, crossing across societies and ages.

Felines in Old Egyptian Fighting

1.1 Consecrated Defenders of Pharaohs

In old Egypt, felines held a consecrated status and were related with the goddess Bastet, the defender of home and family. As the Egyptians participated in military undertakings, felines became venerated colleagues as well as images of security on the war zone. We investigate the social and strict meaning of felines in antiquated Egyptian fighting.

1.2 Vermin Control in Silos and Camps

Common sense met imagery as felines assumed a double part in old Egyptian military settings. Their presence in silos and military camps filled the down to earth need of controlling rat populaces, while their emblematic association with heavenly nature added a layer of persona and importance to the tactical missions. We dive into the archived occasions of felines in military strategies and their more extensive jobs in old Egyptian culture.

Felines in Antiquated Greek and Roman Militaries

2.1 Buddies in Armies and Phalanxes

In antiquated Greece and Rome, felines filled in as allies to officers, especially on lengthy missions and ocean journeys. Their presence in military camps brought both reasonable advantages, for example, bug control, and daily encouragement for troopers not even close to home. We investigate how felines became vital to the regular routines of old champions, cultivating a remarkable fellowship among cats and warriors.

2.2 Emblematic Importance in Folklore

Felines tracked down their direction into the legends of old Greece and Rome, epitomizing both positive and negative characteristics. We analyze how felines were depicted in fantasies and legends, like the relationship with the goddess Artemis in Greek folklore and their apparent association with the otherworldly in Roman convictions. The representative jobs of felines in forming military culture are investigated in this part.

Felines in Archaic European Fighting

3.1 Partners in Palaces and Keeps

During the archaic period, felines became apparatuses in the palaces and keeps of European respectability. Filling in as the two allies to masters and women and bug regulators in the frequently rodent pervaded conditions, felines assumed useful parts in the day to day routines of those participated in archaic fighting. We investigate the homegrown and military elements that elaborate felines in middle age European settings.

3.2 Felines in Attack Fighting

With regards to attack fighting, felines took on unforeseen jobs. From offering close to home help to warriors during delayed attacks to being utilized as covert couriers, felines wound up laced in the special difficulties of middle age military missions. We dive into authentic records and stories that feature the flexibility and creativity of felines during seasons of contention.

Felines in East Asian Fighting

4.1 Imagery and Odd notion in Chinese Fighting

In old China, felines held representative importance in military settings. Accepted to have supernatural characteristics and defensive energies, felines were related with favorable luck and were many times portrayed in craftsmanship and writing connected with fighting. We investigate what the imagery of felines meant for the mentality of Chinese fighters and commandants.

4.2 Samurai and Felines in Primitive Japan

In primitive Japan, felines were venerated for their beauty and spryness, attributes that reverberated with the samurai code. We look at the social and imaginative portrayals of felines in samurai fables and their presence in military families. From friendship to imagery, felines made a permanent imprint on the military customs of primitive Japan.

Felines in Attack Fighting

5.1 The Job of Felines in Archaic Attack Strategies

Attack fighting introduced special difficulties, and felines assumed unforeseen parts in the systems utilized by the two assailants and safeguards. From the viable utilization of felines to ship messages across adversary lines to the notions encompassing their presence during attacks, we dive into the nuanced manners by which felines were woven into the texture of middle age attack fighting.

5.2 The Launch Danger

In a strange and frequently diverting development, verifiable records recount felines being weaponized during attacks. The catapulting of flaring or unhealthy creatures, including felines, was a strategy utilized to spread dread and infection among block-aded populaces. We investigate the whimsical and morally questionable utilization of felines as a feature of attack fighting.

Felines as Emblematic Elements

6.1 Heraldry and Military Images

Felines tracked down their direction into heraldic images and military seals, address-ing a scope of characteristics from spryness to savagery. We investigate how felines were taken on as images by different military elements, reflecting both functional affiliations and a more profound social importance joined to these perplexing animals.

6.2 Felines in Middle age Writing and Craftsmanship

Middle age writing and craftsmanship frequently consolidated felines as images, with their portrayals reaching out past the pragmatic jobs they played in fighting. From enlightened compositions to epic sonnets, we dive into the manners by which felines were portrayed and deciphered in the creative and artistic scenes of middle age Europe.

Fables and Odd notions

7.1 Felines as Signs and Prophets

Odd notions encompassing felines in fighting were common across societies. We investigate the conviction frameworks that credited prophetic capacities to felines, with their ways of behaving and appearances remembered to anticipate triumphs, overcomes, or looming risk on the combat zone.

7.2 Felines and Witches: The Crossing point of Fighting and the Powerful

The relationship among felines and witches during the archaic period had suggestions for impression of cats in military settings. We analyze what strange notions about witches' familiars meant for the treatment of felines in wartime, adding to their double jobs as the two buddies and objects of doubt.

Felines as Mascots and Motivations

8.1 Motivations for Fighters: The Feline as a Military Dream

Felines filled in as dreams for fighters and military pioneers, moving stories of valor and ability. We investigate how the qualities of felines, like dexterity and sharp detects, became analogies for military ideals in the writing and legend of different antiquated and archaic societies.

8.2 Felines as Regimental Mascots

In later military history, felines play assumed the part of regimental mascots, addressing military units and typifying their soul. We dive into instances of popular felines that became darling mascots, adding to the spirit and fellowship of troopers during seasons of war.

Felines and Customs of War

9.1 Feline Penances and Customs

In specific old and archaic societies, felines were exposed to conciliatory customs as a component of war-related services. We investigate the authentic occurrences of feline penances and the social convictions that supported these practices.

9.2 Ceremonial Purposes in Attack Fighting

Attack fighting likewise seen formal purposes of felines, going from their consideration in emblematic functions to acts accepted to guarantee progress in military undertakings. We analyze the convergence of ceremonial practices and the jobs doled out to felines with regards to attack fighting.

7.3 The Cat Corps: Cats in Modern Military History

The records of military history are not just loaded up with stories of human valor and key splendor yet in addition incorporate the frequently disregarded accounts of cat associates who assumed startling parts in the battlefield. This far reaching investigation digs into the charming and various jobs of felines in present day military history, crossing clashes from The Second Great War to contemporary times. From transport felines and channel allies to mascots and treatment creatures, the catlike presence in the tactical scene is a one of a kind and frequently endearing part of the human-creature bond.

The Channel Associates of The Second Great War

1.1 Felines Down and dirty: Quiet Confidants

The Second Great War, set apart by the dreary truth of close quarters conflict, considered the development of felines to be quiet allies to

fighters on the cutting edges.

Their presence gave comfort, friendship, and a passing feeling of predictability in the midst of the repulsions of war. We investigate the individual records and photos that catch the bond produced among troopers and their catlike friends down and dirty.

1.2 Rodents, Channels, and the Feline Arrangement

The channels were overflowing with rodents, presenting serious dangers to the wellbeing and spirit of warriors. Felines, with their regular nature for hunting, became basic partners in the fight against rodents. We dig into the essential arrangement of felines to control vermin populaces and the effect this had on the general prosperity of troops.

Maritime Felines: Watchmen of the Great Oceans

2.1 Felines On board Warships: Keeping up with Cleanliness and Assurance

Yet again as maritime powers extended their armadas during the twentieth 100 years, felines ended up amidst sea fighting. Their jobs stretched out past irritation control to becoming darling mascots, giving friendship and a feeling of solace to mariners confronting the difficulties of life adrift. We investigate the recorded occurrences of maritime felines and their importance in keeping up with cleanliness and lifting everyone's spirits.

2.2 The Ocean Feline Custom: Shipboard Legends

The ocean feline custom, established in sea legends, saw felines viewed as images of best of luck and security against pernicious spirits. Mariners accepted that having a feline on board guaranteed safe journeys and good breezes. We unwind the notions and stories encompassing shipboard felines that became unbelievable figures in maritime history.

Felines in The Second Great War

3.1 Felines on the Home Front: Regular citizen Mates

The Second Great War achieved changes on the home front, and felines assumed parts past the tactical circle. They became soothing allies for families persevering through wartime difficulties, giving a wellspring of comfort and steadiness during fierce times. We investigate the complex jobs of felines on the home front and their effect on regular citizen confidence.

3.2 The Wartime Undertakings of Boat Felines

Maritime vessels kept on having felines on board during The Second Great War, with their jobs growing to incorporate friendship, bug control, and in any event, filling in as mascots for whole boat teams.

The wartime experiences of boat felines, including their endurance in the midst of maritime commitment, add a layer of flexibility to their verifiable story.

Military Working Felines

4.1 Nuisance Regulators on Army installations

In current military settings, felines keep on serving reasonable jobs as irritation regulators on army installations. Their capacity to control rat populaces adds to keeping up with cleanliness and saving assets. We investigate how military functioning felines have become overlooked yet truly great individuals in the fight against bugs, guaranteeing the prosperity of work force and hardware.

4.2 Sent Felines: Cat Mates in Struggle Zones

The presence of felines in struggle zones has been recorded in later military tasks. Felines act as allies to fighters, offering snapshots of rest and an association with a world past the brutal real factors of war. We dive into the accounts of sent felines and their parts in giving solace and kinship in testing conditions.

The Essential Worth of Felines in Fighting

5.1 Felines as Eccentric Partners

While not prepared for battle, felines have exhibited a novel arrangement of abilities that demonstrated beneficial in specific military circumstances. From their sharp faculties to their capacity to cross troublesome territory, we investigate cases where felines became flighty partners, adding to the progress of military activities.

5.2 Felines as Sentinel Gatekeepers

Felines, with their intense faculties, have gone about as sentinel gatekeepers in military conditions. Their capacity to identify moving toward dangers, whether human or creature, has been perceived and used by military faculty. We analyze the essential worth of felines as quiet gatekeepers in different military settings.

Treatment Felines and Consistent reassurance

6.1 Felines in Military Treatment Projects

Perceiving the positive effect of creatures on psychological wellness, treatment felines have turned into an essential piece of military projects pointed toward offering close to home help to support individuals. We investigate the job of felines in restorative mediations, offering solace and friendship to those managing the burdens of military life.

6.2 Cat Friendship for Veterans

Past dynamic assistance, felines play played parts in giving friendship and consistent encouragement to veterans. The quieting presence of felines has been outfit to support the progress to regular citizen life and adapt to the difficulties of post-horrible pressure issue (PTSD). We look at the drives that include felines in supporting veterans' prosperity.

Military Felines as Mascots

7.1 Regimental Mascots: Lifting everyone's spirits

Military units all over the planet have taken on felines as regimental mascots, representing the soul and personality of their separate associations. These mascots become adored figures, partaking in services, marches, and in any event, sending with troops. We investigate the social meaning of military felines as mascots and their part in lifting unit feeling of confidence.

7.2 Outstanding Military Feline Mascots

All through present day military history, certain felines have acquired reputation as famous mascots. From "Leader Feline" on a Regal Naval force submarine to "Sergeant Significant Stubbles" in the U.S. Marine Corps, we dig into the narratives of prominent military feline mascots and their getting through influence on military practices.

The Moral Contemplations of Military Felines

8.1 Creature Government assistance and Moral Treatment

The utilization of felines in military settings raises moral contemplations connected with creature government assistance. We investigate the moral ramifications of including felines in different military jobs, from working felines on bases to mascots taking part in open occasions.

8.2 Support and Guidelines

Basic entitlements activists and associations advocate for the sympathetic treatment of creatures in all specific circumstances, including the military. We look at the administrative systems and rules that address the moral contemplations of including felines in military exercises and the continuous endeavors to guarantee the prosperity of creatures in such conditions.

Felines in War Writing and Craftsmanship

9.1 Artistic Portrayals of Military Felines

The presence of felines in current conflict writing mirrors the complicated connections among fighters and their catlike sidekicks. We investigate how creators have depicted felines in books, brief tales, and verse, catching the close to home bonds shaped amidst struggle.

9.2 Felines in War Movies and Imaginative Portrayals

War films and imaginative portrayals frequently remember felines as representative or strict components for their accounts. From depicting the difficulties of battle to representing trust and strength, we look at how felines have been portrayed in clear line of sight and realistic expressions as impressions of the human involvement with seasons of contention.

The Getting through Tradition of Military Felines

10.1 Respecting Military Felines: Dedications and Accolades

In acknowledgment of the persevering through commitments of military felines, different dedications and accolades have been laid out to respect their heritage. From sculptures to dedicatory plaques, these commemorations commend the extraordinary connection among officers and their catlike companions.

10.2 Military Felines in Mainstream society

The tradition of military felines lives on in mainstream society, with references showing up in movies, writing, and, surprisingly, web-based entertainment. From fictionalized records to narratives, we investigate how military felines keep on catching the creative mind of crowds and keep up with their status as charming figures in the tactical account.

Chapter 8

Cat Symbolism And Superstitions

Felines, with their confounding nature and effortless attitude, have enamored human creative mind for quite a long time, turning into the subjects of rich imagery and strange notions across societies. This far reaching investigation digs into the diverse universe of feline imagery and the frequently interesting, some of the time incongruous, notions related with these catlike animals. From old human advancements to current times, felines have been venerated, dreaded, and beguiled, making a permanent imprint on old stories, writing, workmanship, and ordinary convictions.

Felines in Antiquated Egypt: Holy Watchmen

1.1 Bastet, the Catlike Goddess

In old Egypt, felines held a loved status, with the goddess Bastet representing cat characteristics. Bastet was the goddess of home, fruitfulness, and security, frequently portrayed with the top of a lioness or homegrown feline. We investigate the profound association among felines and Egyptian otherworldliness, where cats were viewed as watchmen and carriers of favorable luck.

1.2 Felines as Defenders of the Home and Then some

In Egyptian families, felines were appreciated buddies as well as adored for their defensive characteristics. Felines were accepted to defend homes from malicious spirits and negative energies. We dig into the customs and convictions encompassing the presence of felines in old Egyptian homes and their jobs in guaranteeing the prosperity of the family.

Middle age Europe: From Witches' Familiars to Images of Karma

2.1 Felines and Witches: The Disrupting Affiliation

In middle age Europe, felines became trapped in the complicated snare of odd notions encompassing black magic. Dark felines, specifically, were accepted to be witches' familiars, prompting their abuse during the witch preliminaries. We investigate how felines went from being defenders to objects of doubt and dread in the archaic European creative mind.

2.2 Fortunate Felines and Old stories

Notwithstanding the relationship with black magic, felines likewise became images of karma and thriving in middle age European old stories.

Stories of felines bringing favorable luck, particularly dark felines, arose close by the notions of their otherworldly associations. We disentangle the division of feline imagery in archaic Europe, where they were both dreaded and worshipped.

Asian Societies: Images of Fortune and Magic

3.1 Maneki-neko: The Coaxing Feline

In different Asian societies, the Maneki-neko, or coaxing feline, is a typical charm accepted to bring favorable luck and success. We investigate the beginnings of the Maneki-neko and its importance in Japanese, Chinese, and other East Asian customs. From paw positions to variety imagery, the Maneki-neko has turned into a social symbol.

3.2 Felines in Chinese Fables: Gatekeepers and Shape-shifters

In Chinese fables, felines are related with both security and supernatural quality. Stories of feline spirits, gatekeepers of homes, and shape-moving animals add to a rich embroidery of cat imagery. We dive into the social subtleties of feline convictions in Chinese folklore and their getting through presence in conventional traditions.

Felines in Norse Folklore: Freyja's Mates

4.1 Freyja and Her Chariot-Pulling Felines

In Norse folklore, the goddess Freyja, related with adoration, richness, and war, is frequently portrayed with chariot-pulling felines. These legendary cats, known as Skogkatts, were venerated as hallowed mates. We investigate the job of felines in Norse folklore and their emblematic association with the strong goddess Freyja.

4.2 Tails of Change: Felines as Shape-shifters

Norse legends highlights stories of felines as shape-shifters, equipped for changing into people or different animals. These shape-moving capacities added a component of secret and sorcery to the imagery of felines in Norse culture. We reveal the accounts that depict felines as subtle and groundbreaking creatures in Norse folklore.

Felines in Islamic Societies: Blended Imagery

5.1 Felines in Islamic Workmanship and Writing

In Islamic societies, felines have held a nuanced place in workmanship and writing. Portrayed in original copies and verse, felines are frequently connected with tidiness and family life. We investigate the blended imagery of felines in Islamic practices, where they are both loved for their class and seen with alert because of their meander-ing nature.

5.2 The Prophet's Affection for Felines

Islamic customs likewise relate cases of Prophet Muhammad's fondness for felines. The Prophet's accounts of generosity toward cats have affected the treatment of felines in Islamic families. We look at the lessons and accounts that exhibit the merciful connection between the Prophet and felines in Islamic culture.

Felines in Japanese Legends: Bakeneko and Nekomata

6.1 Heavenly Felines: Bakeneko and Nekomata

Japanese legends presents heavenly feline animals known as Bakeneko and Nekomata. These otherworldly creatures are accepted to have supernatural abilities, including the capacity to shape-shift and speak with the soul world. We dig into the fables encompassing these extraordinary felines and their jobs in Japanese folklore.

6.2 Feline Sanctuaries and the Love for Cats

Japan's social appreciation for felines reaches out to feline sanctuaries, where cat divinities are loved and worshipped. The Gotoku-ji Sanctuary, specifically, is eminent for its relationship with the Maneki-neko. We investigate the social meaning of feline sanctuaries in Japan and their job in cultivating an association among people and felines.

Dark Felines: Signs of Good and Misfortune

7.1 Dark Felines in Western Notions

Dark felines have for some time been focal figures in Western notions, representing both great and misfortune. We disentangle the problematic convictions encompassing dark felines, from their relationship with witches to their jobs as carriers of thriving and positive signs.

7.2 Dark Felines in Fables and Mainstream society

Dark felines have left their paw prints on fables, writing, and mainstream society. From Edgar Allan Poe's stories to strange notions about running into each other, we investigate the getting through persona of dark felines and their unavoidable presence in the aggregate creative mind.

Felines in Current Imagery

8.1 Felines in Writing and Craftsmanship

In current writing and craftsmanship, felines keep on being strong images, addressing a scope of subjects from freedom and secret to erotic nature and enchantment. We investigate how contemporary creators and craftsmen have utilized feline imagery to convey nuanced implications in their works.

8.2 Web Felines and Mainstream society

The ascent of web felines, from viral images to online entertainment sensations, has made another element of feline imagery in mainstream society. Felines like Testy Feline and Lil Buddy have become notable figures, affecting patterns and leaving an enduring effect on the computerized scene. We analyze the job of web felines in molding present day feline imagery.

The Notions That Persevere

9.1 Feline Intersection: Signs and Society Convictions

The notion of a dark feline intersection one's way endures in different societies, each with its own translations and convictions. We investigate the beginnings of this notion and the varieties in people convictions in regards to feline intersections.

9.2 Felines and the Powerful: Spirits and Signs

Odd notions about felines reach out to their apparent associations with the power-ful. From identifying spirits to filling in as signs, we dive into the notions that depict felines as otherworldly creatures with capacities past the common.

8.1 Cats and Witchcraft

The convergence of felines and black magic has a long and complex history, interweaved with old stories, strange notions, and social insights. Felines, frequently connected with secret and freedom, became focal figures in the witch preliminaries of archaic Europe, acquiring both worship and dread. This exhaustive investigation digs into the multi-layered connection among felines and black magic, crossing different societies, verifiable periods, and the persevering through influence on well known creative mind.

Felines in Old Folklore and Agnostic Convictions

1.1 Consecrated Cats and Agnostic Divinities

In old folklore, felines were frequently connected with divinities and loved as sacrosanct creatures. From the Egyptian goddess Bastet to Norse fables highlighting Freyja's chariot-pulling felines, we investigate the positive and loved jobs that felines played in old agnostic convictions.

1.2 Felines as Familiars in Antiquated Supernatural Practices

The idea of familiars, otherworldly substances supporting witches in their en-chanted practices, can be followed back to antiquated times. Felines were accepted to have supernatural characteristics and were viewed as ideal allies for people rehearsing different types of magic. We dive into the antiquated underlying foundations of felines as familiars and their apparent association with otherworldly domains.

Middle age Europe: Felines, Witches, and the Probe

2.1 Dark Felines and the Witches' Recognizable Generalization

The middle age time frame saw an exceptional change in the impression of felines, especially dark felines. As Christianity acquired conspicuousness, felines, and particu-larly dark ones, became related with black magic and Satan. We investigate how the generalization of the witches' natural arose and added to the boundless apprehension about felines.

2.2 Felines as Accessories: The Job of Felines in Witch Preliminaries

Felines ended up entrapped in the witch preliminaries that cleared across Europe. Blamed for being witches' associates, felines were oppressed, tormented, and killed close by those blamed for rehearsing black magic. We dig into verifiable records and analyze the appalling destiny of felines during this dull part ever.

The Imagery of Felines in Black magic

3.1 Felines as Images of Change

In black magic, felines have been images of change and shape-moving. Old stories and enchanted customs frequently portray witches assuming the type of felines to move covertly or travel between domains. We investigate the imagery of felines with regards to supernatural changes and their relationship with the magical parts of black magic.

3.2 Felines as Watchmen of Mystical Spaces

Felines are frequently viewed as watchmen of mystical spaces in different black magic practices. Their presence is accepted to improve the energy of sacrosanct places and proposition security to professionals. We look at the imagery of felines as supernatural watchmen and their job in establishing and keeping up with mysterious conditions.

Witches' Familiars: The Otherworldly Association

4.1 The Idea of Familiars in Black magic

The idea of familiars, profound substances accepted to help witches in their otherworldly functions, has been a common subject in black magic legend. Felines, close by different creatures, have been viewed as expected familiars, shaping an otherworldly association with professionals. We investigate the authentic and social parts of familiars in black magic.

4.2 Felines as Mysterious Aides and Buddies

In different black magic customs, felines are seen as mysterious aides and buddies, helping professionals in their otherworldly excursion. From divination to spellwork, we dig into how felines are seen as natural and profoundly adjusted creatures, improving the supernatural acts of witches.

Felines in People Enchantment and Tricky Society Customs

5.1 Felines in People Mending and Divination

Society enchantment and sly people customs frequently integrated felines into recuperating practices and divination ceremonies. Felines were accepted to have exceptional characteristics that could be saddled for mysterious purposes. We investigate the jobs of felines in society sorcery and their importance in customary recuperating rehearses.

5.2 Sly People and Feline Wizardry

Sly people, people rehearsing society wizardry and offering administrations to their networks, frequently had relationship with felines in their supernatural undertakings. We look at how felines were coordinated into the acts of clever people, going about as mysterious partners chasing positive results.

Present day Black magic and the Feline Natural

6.1 Felines in Contemporary Black magic

Present day black magic, frequently alluded to as Wicca and other agnostic customs, keeps a positive perspective on felines as enchanted creatures and familiars. We investigate how contemporary witches embrace felines as buddies and profound partners in their otherworldly operations, encouraging an association with old customs.

6.2 Feline Familiars in Wiccan Practices

Wicca, a cutting edge agnostic religion, puts importance on the idea of familiars, with felines being among the inclined toward decisions. We dig into Wiccan works on including feline familiars, the ceremonies related with their determination, and the profound otherworldly bond fashioned among witches and their catlike buddies.

Felines and Mending Enchantment

7.1 Felines as Mending Partners

In different mysterious customs, felines are related with mending and health. Their quieting presence and instinctive nature make them ideal allies for people working on mending sorcery. We investigate how felines have been coordinated into customs and practices zeroed in on physical and otherworldly prosperity.

7.2 Feline Wizardry for Insurance and Warding

Felines are much of the time thought about mystical defenders, and their pictures or portrayals are utilized in defensive enchantment and warding ceremonies. From charms to spellwork, we look at how felines assume a part in mystical practices pointed toward shielding people and spaces.

Strange notions Encompassing Felines in Black magic

8.1 Dark Felines: Signs of Good and Misfortune

Strange notions encompassing dark felines continue in present day times, with convictions changing from one locale to another. We investigate the polarity of dark felines being viewed as the two signs of best of luck and carriers of misfortune, analyzing the getting through effect of these notions.

8.2 Other Feline Notions in Black magic

Past dark, different notions are related with felines in black magic. From the meaning of a feline intersection one's way to the confidence in felines as shape-moved witches, we dive into the notions that keep on forming impression of felines in other-worldly settings.

The Feline Goddess Recovery

9.1 Contemporary Recovery of Feline Goddess Love

Lately, there has been a restoration of interest in old feline goddesses and their love. Present day specialists investigate the profound parts of feline goddesses like Bastet, looking for motivation from old customs. We look at the contemporary recovery of feline goddess love and its importance in current otherworldly practices.

9.2 Feline Goddesses in Mainstream society

The impact of feline goddesses stretches out to mainstream society, where references to Bastet and other cat divinities can be tracked down in writing, workmanship, and media. We investigate how feline goddesses have become images of female power and persona in contemporary social articulations.

Felines in Writing, Film, and Mainstream society

10.1 Felines as Enchanted Creatures in Writing

Writing has frequently depicted felines as enchanted creatures, whether as familiars, shape-shifters, or gatekeepers of mysterious domains. We investigate how creators have mixed mysterious characteristics into artistic cat characters, adding to the getting through relationship among felines and black magic.

10.2 Felines in Film and TV: From Familiars to Witches

The depiction of felines in film and TV has propagated the connection among felines and black magic. Whether as familiars, sidekicks, or supernatural creatures by their own doing, felines have made a permanent imprint on mainstream society. We analyze notable cat characters and their portrayal in the enchanted domain of diversion.

8.2Black Cats: Superstitions and Symbolism

Dark felines, with their smooth covers and entrancing eyes, have been at the focal point of an embroidery woven with strange notions and imagery for quite a long time. From the beginning of time, these catlike colleagues have been both respected and dreaded, epitomizing a perplexing duality in the human creative mind. This exhaustive investigation dives into the complex universe of dark feline notions and imagery, spreading over different societies, authentic periods, and the getting through influence on prevalent views.

Old Egypt - The Loved Bastet
1.1 Bastet: Goddess of Home and Insurance

In old Egypt, dark felines were related with the goddess Bastet, a god addressing home, ripeness, and security. Bastet, frequently portrayed with the top of a lioness or homegrown feline, raised the situation with dark felines to that of venerated gatekeepers. We investigate the basic job dark felines played in old Egyptian otherworldliness and their emblematic association with the heavenly.

1.2 Felines as Sacrosanct Partners

Dark felines in antiquated Egypt were not only pets; they were viewed as hallowed sidekicks. Their presence in homes was accepted to bring favors, and they were adored for their capacity to avoid malicious spirits. We dig into the meaning of dark felines as defenders and images of favorable luck in the rich embroidery of antiquated Egyptian convictions.

Middle age Europe - From Worship to Mistreatment
2.1 Felines and Black magic: The Development of Dread

As Europe progressed into the middle age time frame, the impression of dark felines went through a sensational shift. With the ascent of Christianity, these once-adored animals became ensnared in the developing trepidation and doubt encompassing black magic. Dark felines, specifically, were accepted to be witches' familiars, prompting their mistreatment and relationship with dim enchantment.

2.2 Dark Felines as Signs: Strange notions Flourish

Strange notions about dark felines increased during bygone eras, and they became signs of setback. Running into a dark feline was considered unfortunate, and the feeling of dread toward these catlike animals penetrated society. We investigate the foundations of these notions and their getting through influence on the social mind.

Asian Societies - The Yin and Yang of Dark Felines
3.1 Dark Felines in Japanese Old stories

In Japanese old stories, dark felines have a nuanced presence, exemplifying both positive and negative imagery. From warding off abhorrent spirits to being related with shape-moving powerful elements, we unwind the intricacies of dark feline imagery in Japanese culture.

3.2 Chinese Imagery: Karma and Assurance

Chinese culture, then again, frequently sees dark felines as images of karma and insurance. Dark is related with warding off insidiousness, and dark felines are accepted to bring favorable luck. We dig into the differentiating view of dark felines in Asian societies and their jobs in society convictions.

Dark Felines in Norse Folklore - Freyja's Colleagues

4.1 Freyja and the Chariot-Pulling Felines

In Norse folklore, dark felines are connected to the goddess Freyja, a strong figure related with adoration, fruitfulness, and war. Freyja's chariot-pulling felines, frequently portrayed as skogkatts, hold a unique spot in Norse old stories. We investigate the meaning of dark felines as hallowed allies to the venerated goddess.

4.2 Shape-Moving Cats: Felines in Norse Legends

Norse legends presents shape-moving felines, fit for changing into different animals or even people. Dark felines, with their mysterious characteristics, become creatures of secret and wizardry in Norse folklore. We disentangle the stories that paint dark felines as subtle and groundbreaking substances.

Dark Felines in Contemporary Wicca and Black magic

5.1 Dark Felines as Familiars

In contemporary Wicca and black magic, dark felines recover their status as mystical creatures and familiars. A long way from being images of mishap, these catlike mates are embraced for their secretive and natural characteristics. We investigate how dark felines are coordinated into present day supernatural practices and the imagery they hold in contemporary black magic.

5.2 The Positive Imagery of Dark Felines

Wiccan and present day black magic practices frequently view dark felines as images of security and clairvoyant energy. Their relationship with the inconspicuous and the baffling lines up with the otherworldly parts of these profound practices. We dig into the positive imagery ascribed to dark felines in the assorted scene of contemporary black magic.

Fables and Mainstream society - Dark Felines in Writing and Film

6.1 Dark Felines in Writing: Imagery and Analogy

In writing, dark felines have been utilized as images and illustrations, addressing a scope of subjects from secret and change to strange notion. We investigate how creators have utilized dark felines to pass nuanced implications and add layers of intricacy on to their stories.

6.2 Dark Felines in Film: From Odd notion to Natural

The depiction of dark felines in film has propagated the two notions and positive affiliations. Whether as unfavorable signs or mysterious familiars, dark felines have made a permanent imprint on artistic narrating. We analyze famous dark feline characters and their portrayal in the mysterious domain of diversion.

Dark Felines and Halloween - An Eccentric Association

7.1 Dark Felines and Halloween Old stories

Halloween, with its foundations in antiquated Celtic customs, has for some time been related with odd notions encompassing dark felines. From witches' familiars to images of the extraordinary, dark felines assume an unmistakable part in Halloween old stories. We investigate the persevering through association between dark felines and the spookiest of occasions.

7.2 The Dark Feline Paradigm in Halloween Symbolism

The dark feline prime example includes noticeably in Halloween symbolism, encapsulating both trepidation and interest. From enhancements to outfits, the presence of dark felines during Halloween adds to the puzzling and mysterious air of the time. We dive into the imagery and odd notions encompassing dark felines during this bubbly time.

Dark Felines in Notions - Signs of Good and Misfortune

8.1 Running into a Dark Feline

The notion of running into a dark feline has endured through the ages, with fluctuating translations and convictions. While certain societies see it as an indication of looming setback, others view it as a positive sign. We investigate the starting points of this notion and its social varieties.

8.2 Dark Felines on Friday the thirteenth

Friday the thirteenth, frequently viewed as an unfortunate day, turns out to be much more inauspicious when dark felines are involved. The notion encompassing dark felines on this day adds an additional layer of secret and dread.

We look at the conversion of strange notions that increase the imagery of dark felines on Friday the thirteenth.

Dark Felines as Pet Partners - Breaking Generalizations

9.1 Embracing Dark Felines: Dispersing Fantasies

In spite of the notions and negative imagery, numerous people decide to take on dark felines as pets. Creature shields frequently face difficulties in tracking down homes for dark felines because of waiting notions. We investigate endeavors to scatter fantasies and advance the reception of dark felines as cherishing and meriting buddies.

9.2 Positive Affiliations: Dark Felines as Recognizable Colleagues

In families all over the planet, dark felines are appreciated as recognizable buddies, breaking liberated from the shadows of odd notions. We dig into endearing accounts of dark felines giving pleasure, solace, and positive energy into the existences of their human sidekicks.

The Social Flexibility of Dark Felines

10.1 Dark Felines in Craftsmanship and Writing Today

Contemporary craftsmen and creators keep on investigating the imagery of dark felines in their work, testing generalizations and commending the appeal of these baffling cats. We look at how dark felines are depicted in current creative articulations and writing, adding to a moving story.

10.2 The Eventual fate of Dark Feline Imagery

As cultural viewpoints develop, the fate of dark feline imagery is set apart by a re-examination of social stories. We investigate how evolving mentalities, schooling, and mindfulness are forming a more certain and nuanced comprehension of dark felines in the shared mindset.

8.3 Cats in Different Cultures and Their Symbolic Meanings

Felines, with their cryptic presence and elegant disposition, have been adored, dreaded, and celebrated across assorted societies over the entire course of time. Past being cherished buddies, felines play played emblematic parts in folklore, fables, and otherworldly convictions. This investigation dives into the multi-layered universe of felines in various societies, disentangling the representative implications credited to these catlike animals across different human advancements and customs.

Felines in Old Egypt - Watchmen of the Heavenly

1.1 Bastet: The Catlike Goddess

In old Egypt, felines held a consecrated status, encapsulated by the goddess Bastet. With the top of a lioness or homegrown feline, Bastet represented home, richness, and insurance. Felines were adored as watchmen of the home, and their presence was accepted to bring gifts and avoid insidious spirits. We investigate the profound association among felines and otherworldliness in old Egyptian culture.

1.2 Felines as Images of Sovereignty and Effortlessness

Notwithstanding their job in strict convictions, felines were related with eminence in old Egypt. The pharaohs esteemed felines for their beauty and balance, frequently integrating cat themes into workmanship and engineering. The emblematic meaning of felines as magnificent and divine creatures pervaded Egyptian culture, making a permanent imprint on their social heritage.

Felines in Japanese Legends - Supernatural Watchmen and Shape-shifters

2.1 Maneki-neko: The Enticing Feline

Japanese culture embraces the imagery of felines in the famous charm known as Maneki-neko, or the enticing feline. Frequently portrayed with a raised paw, the Maneki-neko is accepted to bring favorable luck and thriving. We investigate the beginnings and social meaning of this notable cat image in Japanese fables.

2.2 Bakeneko and Nekomata: Heavenly Feline Creatures

Japanese legends presents otherworldly feline animals known as Bakeneko and Nekomata. These otherworldly creatures are accepted to have mystical abilities, including shape-moving and correspondence with the soul world. We dive into the old stories encompassing these puzzling feline elements and their jobs in Japanese folklore.

Felines in Norse Folklore - Freyja's Chariot-Pulling Friends
3.1 Freyja and Her Skogkatts
Norse folklore includes the goddess Freyja, related with affection, richness, and war, frequently portrayed with chariot-pulling felines. These legendary cats, known as Skogkatts, were venerated as consecrated allies to Freyja. We investigate the imagery of felines in Norse folklore and their representative association with this strong goddess.

3.2 Felines as Shape-shifters and Representative Substances
Norse legends adds one more layer to the imagery of felines by depicting them as shape-shifters. Felines were accepted to can change into people or different animals, adding a component of secret and sorcery to their representative importance.

We disentangle the stories that portray felines as slippery and extraordinary creatures in Norse culture.

Felines in Chinese Culture - Gatekeepers of Favorable luck
4.1 Felines as Images of Karma and Flourishing
Chinese culture has an uplifting outlook of felines as images of karma and thriving. Dark, specifically, is related with warding off evil, making dark felines particularly promising. From legends to customary workmanship, we investigate how felines represent favorable luck and positive energy in Chinese practices.

4.2 The Five Endowments: Felines in Feng Shui
In Feng Shui, the old Chinese act of orchestrating with the climate, felines are viewed as bearers of the five favors: life span, riches, wellbeing, temperance, and a quiet demise. We dig into the job of felines in Feng Shui and their importance in making an amicable living space.

Felines in Islamic Societies - Images of Tidiness and Family life
5.1 Felines in Islamic Workmanship and Writing
Islamic societies have portrayed felines in craftsmanship and writing, frequently connecting them with neatness and home life. Felines are highlighted in compositions and verse, displaying their agile presence. We investigate the blended imagery of felines in Islamic customs, where they are both respected for their tastefulness and seen with alert because of their meandering nature.

5.2 The Prophet's Love for Felines
Islamic customs relate cases of Prophet Muhammad's friendship for felines, featuring their positive job in the family. The Prophet's accounts of generosity toward cats have impacted the treatment of felines in Islamic families. We look at the lessons and stories that feature the merciful connection between the Prophet and felines in Islamic culture.

Felines in Hinduism - The Heavenly Cat Friend
6.1 Goddess Maa Durga's Catlike Mount
In Hinduism, the goddess Maa Durga is in many cases portrayed riding a lion or tiger, underlining the representative association between the catlike and the heavenly.

Felines, in different structures, are related with goddesses and their heavenly properties. We investigate the imagery of felines as consecrated allies to Hindu gods.

6.2 The Ailurophile Divinity: Master Shiva

Ruler Shiva, one of the chief divinities in Hinduism, is depicted as an ailurophile, communicating a profound love for felines.

Felines are viewed as propitious in the love of Master Shiva, and their presence is accepted to bring endowments and security. We dive into the meaning of felines in Hindu strict practices.

Felines in Celtic Legends - Watchmen of the Otherworld

7.1 Cait Sith: Pixie Felines and Shapeshifters

Celtic legends presents the idea of pixie felines known as Cait Sith, accepted to be otherworldly creatures fit for shape-moving. These secretive felines were related with the pixie domain and were accepted to bring both great and horrible luck. We investigate the legends encompassing Cait Sith and their emblematic presence in Celtic practices.

7.2 Felines as Gatekeepers of Homes and Otherworldliness

Celtic societies thought about felines as gatekeepers of homes and profound domains. Their nighttime exercises and sharp faculties made them images of assurance against noxious powers. We disentangle the imagery of felines as enchanted gatekeepers in Celtic fables and their parts in warding off bad energies.

Felines in Local American Societies - Soul Guides and Defenders

8.1 The Job of Felines in Local American Otherworldliness

A few Local American clans view felines as otherworldly creatures and guides. Felines are viewed as defenders, bestowing intelligence and direction to the individuals who interface with their energy. We investigate the assorted manners by which felines are venerated and incorporated into the otherworldly acts of Local American societies.

8.2 Imagery of Wild Felines in Local American Folklore

Wild felines, like the cougar or catamount, hold explicit representative implications in Local American folklore. These catlike animals are frequently connected with strength, dexterity, and the soul of the chase. We dig into the imagery of wild felines in Local American societies and their jobs as profound emblems.

Felines in Caribbean Fables - Jumbies and Otherworldly Gatekeepers

9.1 The Jumbee Feline: Powerful Defenders

In Caribbean fables, the Jumbee Feline is a legendary cat animal accepted to safeguard people from malicious spirits. The Jumbee Feline's powerful characteristics make it an image of guardianship and otherworldly safeguard. We investigate the legends encompassing Jumbee Felines and their representative importance in Caribbean societies.

9.2 Feline Bones and People Sorcery

The utilization of feline bones in Caribbean people sorcery is a training established in imagery and otherworldly convictions. Feline bones are accepted to have enchanted properties, filling in as charms for security and favorable luck. We dive into the social meaning of feline bones in Caribbean society enchantment and their part in profound practices.

Felines in Contemporary Worldwide Culture - Web Sensations and Then some

10.1 Web Felines: From Viral Sensations to Social Symbols

In the contemporary advanced age, felines have become worldwide VIPs through the web. From Testy Feline to Nyan Feline, cat sensations have caught the hearts of millions, impacting mainstream society and making new layers of imagery. We investigate the peculiarity of web felines and their effect on the worldwide impression of cat imagery.

10.2 Felines in Worldwide Workmanship and Style

Felines keep on being unmistakable figures in worldwide craftsmanship and style. Specialists and creators draw motivation from cat style, making a combination of customary imagery and present day feel. We look at how felines are addressed in the contemporary imaginative scene and their job as images of style and elegance.

Chapter 9

Real-Life Heroes And Notorious Cats

The universe of felines is an embroidery woven with accounts of both genuine legends and infamous characters. From the endearing stories of cat fortitude and dedication to the strange and wicked shenanigans of famous felines, this investigation digs into the diverse universe of felines in the chronicles of mankind's set of experiences. Go along with us on an excursion that traverses landmasses and hundreds of years, uncovering the uncommon and some of the time baffling jobs that felines have played in the existences of individuals.

The Valiant Mousers of Boats

1.1 Boat Felines All through Oceanic History

For a really long time, felines have been crucial individuals from oceanic groups, procuring their keep as master mousers. We dive into the job of boat felines, investigating how these catlike legends safeguarded valuable freight from rodents and odd notions the same. From old nautical societies to the Period of Investigation, transport felines have been enduring partners on sea journeys.

1.2 Uncelebrated Yet truly great individuals of the Ocean: Felines in Maritime Fighting

Past their job as mousers, a few felines became startling legends in maritime fighting. Whether filling in as boat mascots or cautioning of looming risk, felines assumed urgent parts in maritime commitment. We investigate the stories of these overlooked yet truly great individuals and their effect on the destinies of mariners and boats.

The Feline That Served Down and dirty

2.1 Felines in Wartime: The Cat Forefront

During seasons of war, felines ended up on the bleeding edges, serving close by fighters down and dirty. Their presence gave solace, friendship, and, surprisingly, upper hands. We inspect the accounts of wartime felines and their remarkable commitments during clashes across the hundreds of years.

2.2 Simon the Feline: A Decoration of Boldness

The tale of Simon, a feline who served on the HMS Amethyst during the Yangtze Episode in 1949, stands apart as a demonstration of cat boldness. Granted the Dickin Decoration, what might be compared to the Victoria Cross, Simon turned into an image of boldness and flexibility. We dive into the uncommon story of Simon and his effect on maritime history.

Feline Robbers and Gem Criminals

3.1 The Captivating Universe of Feline Thieves

Not all catlike reputation comes from the combat zone. A few felines procured their place in history as famous feline criminals. We investigate the stories of felines with a propensity for taking everything from sparkling items to ordinary things, leaving a path of secret and entertainment.

3.2 Fur-tive Heists: Famous Felines in the Criminal Hidden world

From gem burglaries to historical center heists, a few felines accomplished reputation for their criminal capers. Their finesse and covertness made them captivating figures in the criminal hidden world. We disentangle the tales of these famous cats who became legends in the archives of feline thievery.

Feline Big names of the Cinema

4.1 Hollywood's Catlike Stars

Felines have graced the cinema as magnetic and some of the time devilish characters. From works of art like "Breakfast at Tiffany's" to vivified sensations like "Puss in Boots," we investigate the effect of cat VIPs on the universe of film. These felines brought appeal, humor, and a dash of wizardry to realistic narrating.

4.2 Web Sensations: From Viral Recordings to Worldwide Acclaim

In the advanced age, felines have become web sensations, dazzling crowds all over the planet. Images, viral recordings, and virtual entertainment stages have transformed common felines into worldwide famous people. We analyze the ascent of web felines and their effect on mainstream society, from Surly Feline to Nyan Feline.

The Secretive Universe of Feline Sanctuaries

5.1 Felines as Hallowed Creatures: The Beginnings of Feline Sanctuaries

In different societies, felines are viewed as hallowed creatures, and feline sanctuaries have arisen as spots of love and respect. We investigate the starting points of feline sanctuaries, from antiquated civilizations to current asylums devoted to the catlike divinities. These sanctuaries act as a demonstration of the getting through profound association among people and felines.

5.2 The Perplexing Maneki-neko: Fortune and Success

Quite possibly of the most notorious image in feline sanctuary culture is the Maneki-neko, the coaxing feline. We disentangle the secret behind this catlike doll, investigating its beginnings, imagery, and the convictions encompassing its capacity to bring fortune and thriving. From Japan to the worldwide stage, the Maneki-neko has turned into an image of best of luck and appeal.

Feline Imagery in Old stories and Strange notions

6.1 Dark Felines: Signs of Good and Misfortune

Dark felines have for quite some time been focal figures in old stories and odd notions, representing both great and misfortune. We disentangle the problematic convictions encompassing dark felines, from their relationship with witches to their jobs as carriers of success and positive signs.

6.2 Feline Intersection: Signs and People Convictions

The notion of a dark feline intersection one's way continues in different societies, each with its own translations and convictions. We investigate the starting points of this notion and the varieties in people convictions with respect to feline intersections.

Felines and Black magic

7.1 Felines as Familiars in Old Supernatural Practices

The idea of familiars, profound substances helping witches in their otherworldly practices, can be followed back to antiquated times. Felines were accepted to have mysterious characteristics and were viewed as ideal allies for people rehearsing different types of otherworldliness.

7.2 Dark Felines and the Witches' Recognizable Generalization

The middle age time frame saw an exceptional change in the view of felines, especially dark felines. As Christianity acquired noticeable quality, felines, and particularly dark ones, became related with black magic and Satan. We investigate how the generalization of the witches' natural arose and added to the far and wide apprehension about felines.

Felines in Craftsmanship and Writing

8.1 Felines in Antiquated and Renaissance Craftsmanship

All through craftsmanship history, felines have been portrayed in different structures, from strict imagery to pictures of regarded people. We investigate the advancement of feline portrayal in workmanship, analyzing how these catlike animals became dream and subject for prestigious specialists.

8.2 Artistic Felines: From Murmuring Allies to Heroes

In writing, felines have been depicted as secretive sidekicks, supernatural creatures, and even heroes by their own doing. From Edgar Allan Poe's stories to Lewis Carroll's Cheshire Feline, we disentangle the abstract tradition of felines and their getting through effect on narrating.

Felines in Current Fighting

9.1 Felines in Antiquated and Middle age Fighting

Felines play played surprising parts in fighting since forever ago. From old times to archaic fights, felines were utilized for their extraordinary capacities, whether as vermin regulators or wellsprings of solace for troopers. We investigate the less popular accounts of felines in antiquated and middle age fighting.

9.2 The Feline Corps: Felines in Current Military History

In later times, felines have ended up associated with present day military history. From transport felines to mascots in armed force units, felines have given friendship, solace, and once in a while even upper hands. We dive into the entrancing accounts of the Feline Corps and their commitments during seasons of contention.

The Social Effect of Infamous Felines

10.1 Notorious Felines in Mainstream society

Infamous felines have made a permanent imprint on mainstream society, affecting everything from writing and workmanship to images and product. We investigate the social effect of scandalous felines, from their depiction in kid's shows to their status as images of resistance and underhandedness.

10.2 Felines as Symbols of Opposition

A few felines have become images of opposition and unrest, testing cultural standards and motivating developments. We dive into the narratives of felines that have become symbols of opposition, epitomizing the soul of autonomy and resistance.

From Famous to Darling: The Reclamation of Notorious Felines

11.1 Recovery and Reception of Famous Felines

Notwithstanding their famous notorieties, a few felines track down recovery through restoration and reception. Creature government assistance associations and committed people work to furnish infamous felines with another opportunity at a caring home. We investigate the endeavors to restore and rehome felines that stand out enough to be noticed.

11.2 The Tradition of Genuine Legends and Infamous Felines

In the fabulous embroidery of cat history, the tales of genuine legends and famous felines have woven together to make a rich story. From the grit of wartime felines to the baffling universe of feline robbers, these stories mirror the unpredictable connection among people and their catlike partners. As we investigate the tradition of these surprising felines, we gain bits of knowledge into the persevering through influence they have had on our discernments, societies, and hearts.

9.1 Unsung Heroes: Cats in Search and Rescue

In the huge scene of search and salvage tasks, where people and creatures cooperate to save lives, one frequently neglected at this point exceptional classification of legends arises — the pursuit and salvage felines. These catlike specialists on call, with their sharp detects, deftness, and resolute assurance, assume urgent parts in finding missing people, giving solace in the midst of emergency, and adding to the outcome of search and salvage missions. This investigation digs into the phenomenal universe of search and salvage felines, revealing insight into their preparation, achievements, and the significant effect they have on the lives they contact.

The One of a kind Capacities of Felines

1.1 Cat Detects: A Characteristic Benefit

Felines have noteworthy tactile capacities that make them appropriate for search and salvage missions. From intense hearing to outstanding night vision, we investigate how these normal cat characteristics become priceless resources in testing conditions.

The exceptional characteristics of felines establish the groundwork for their job as uncelebrated yet truly great individuals in search and salvage tasks.

1.2 Deftness and Climbing Skill

The deftness and climbing ability of felines make them exceptional to explore different territories, from thick timberlands to metropolitan scenes. Their capacity to get to hard-to-arrive at places demonstrates fundamental in search and salvage situations. We analyze how these actual properties add to the viability of cat search and salvage endeavors.

The Historical backdrop of Felines in Search and Salvage

2.1 Antiquated Roots: Felines as Watchmen and Guides

The utilization of felines in search and salvage activities has roots that stretch back through history. We investigate how antiquated societies perceived the inherent capacities of felines and integrated them into their practices as watchmen and guides. From antiquated human advancements to bygone eras, felines have been quiet sidekicks in the journey for viewing as the lost.

2.2 Current Acknowledgment: The Rise of Formal Preparation

In later times, the proper preparation of search and salvage felines has earned respect. We follow the development of felines from instinctual responders to prepared experts, analyzing the trailblazers who perceived the capability of cat help with search and salvage endeavors.

Preparing Search and Salvage Felines

3.1 Structure on Normal Impulses

While felines have inborn pursuit and salvage capacities, formal preparation improves and refines these senses. Mentors work with cats to foster explicit abilities, including fragrance following, perceiving trouble flags, and exploring testing conditions. We dig into the preparation procedures that change felines into capable hunt and salvage friends.

3.2 The Job of Encouraging feedback

Encouraging feedback assumes an essential part in preparing search and salvage felines. Dissimilar to customary dutifulness preparing, search and salvage preparing centers around making areas of strength for a between the feline and its overseer. We investigate how uplifting feedback procedures are utilized to propel and remunerate felines during instructional meetings.

Search and Salvage Felines All over the Planet

4.1 The US: Spearheading Endeavors

In the US, search and save felines have become basic individuals from calamity reaction groups. From catastrophic events to man-made crises, cat responders add to endeavors in finding survivors. We look at prominent cases and the associations at the front of search and salvage feline organizations.

4.2 Worldwide Commitments: From Japan to Europe

Search and protect felines have put forth huge commitments to catastrophe reaction attempts all over the planet. In Japan, for instance, where tremors and waves present consistent dangers, cat responders assume imperative parts in finding people caught in rubble. Also, European nations have embraced the help of search and salvage felines in different situations.

The Science Behind Aroma Following

5.1 Cat Olfactory Capacities

The olfactory capacities of felines are surprising, with aroma receptors that far outperform those of people. We investigate the science behind cat aroma following, analyzing how felines utilize their noses to recognize fragrances over huge spans and in testing conditions. The job of fragrance in search and salvage missions turns into a urgent part of their commitment.

5.2 Aroma Segregation and Target Recognizable proof

Search and safeguard felines go through preparing to segregate between various aromas, zeroing in on recognizing explicit targets related with human presence. We dive into the complexities of fragrance segregation and how felines are prepared to pinpoint people in tremendous and complex conditions.

Eminent Pursuit and Salvage Feline Examples of overcoming adversity

6.1 Victories in Cataclysmic events

Search and save felines play played essential parts in catastrophic event reactions. From quakes to storms, we investigate examples of overcoming adversity where cat responders had an effect by finding survivors, giving solace, and adding to the general viability of salvage tasks.

6.2 Metropolitan Inquiry and Salvage: Exploring Cityscapes

In metropolitan conditions, search and salvage felines exhibit their deftness and versatility. We analyze situations where cat responders effectively explored cityscapes, finding missing people amidst complex designs and testing territories.

Search and Salvage Felines in Wild Conditions

7.1 Exploring Thick Woodlands and Wild

The wild presents extraordinary difficulties for search and salvage activities, and felines succeed in these conditions. We investigate how cat responders add to finding people in thick timberlands, rugged locales, and other wild settings. Their capacity to move through testing territory turns into a significant resource.

7.2 Wild Endurance: Felines as Buddies

Past their jobs in viewing as the lost, search and protect felines give friendship and solace to people anticipating salvage in wild conditions. We dive into occasions where cat responders have become encouraging signs for those abandoned in far off areas.

The Human-Creature Bond in Search and Salvage

8.1 Daily encouragement and Solace

The presence of search and salvage felines goes past their specialized commitments. These catlike responders offer everyday reassurance and solace to the two heros and people anticipating salvage. We investigate the significant effect of the human-creature bond in the high-stakes climate of search and salvage activities.

8.2 Pressure Decrease and Injury Alleviation

The pressure and injury related with search and protect missions can overpower. Search and protect felines add to pressure decrease and injury help, making a feeling of business as usual and giving an encouraging presence in the midst of mayhem. We look at the restorative job of cat responders in high-pressure circumstances.

Difficulties and Constraints

9.1 Natural Difficulties

Search and salvage activities frequently occur in different and unusual conditions. We investigate the natural difficulties looked via search and salvage felines, from outrageous weather patterns to exploring dangerous territory. Understanding these difficulties is fundamental in refining preparing and organization techniques.

9.2 Public Insight and Acknowledgment

Regardless of their commitments, search and salvage felines frequently face an absence of public acknowledgment contrasted with their canine partners. We analyze the difficulties in bringing issues to light about the capacities of cat responders and the significance of recognizing their jobs in search and salvage endeavors.

Future Points of view and Developments

10.1 Progressions in Preparing Methods

The field of search and salvage is dynamic, with continuous progressions in preparing strategies for cat responders. We investigate imaginative ways to deal with preparing that influence innovation, social science, and cooperative endeavors to upgrade the abilities of search and salvage felines.

10.2 Coordinating Innovation: Robots and Sensors

The joining of innovation, including robots and sensors, improves the capacities of search and salvage felines. We analyze how these mechanical progressions supplement the innate capacities of cat responders, extending their range and viability in finding missing people.

Search and Salvage Felines as Persuasive Figures

11.1 Cat Versatility and Assurance

The accounts of search and salvage felines are not only stories of courage; they typify characteristics of flexibility and assurance. We investigate how these catlike responders move people to conquer difficulties, adjust to their environmental clements, and endure even with affliction.

11.2 Encouraging Sympathy and Compassion

The sympathy and compassion showed via search and salvage felines leave an enduring effect on the human mind. We dig into how these catlike legends encourage a feeling of sympathy, compassion, and interconnectedness, impacting cultural mentalities toward the two creatures and individual people.

9.2 Infamous Cats: Tales of Cat Burglars and Mischievous Felines

In the tremendous embroidered artwork of cat history, a few felines stand apart not so much for their courage or faithfulness but rather for their famous deeds and devilish endeavors. These notorious felines, frequently named as feline robbers, have left their paw prints on the chronicles of mankind's set of experiences, making stories that reach from baffling burglaries to perky jokes. This investigation digs into the captivating universe of notorious felines, unwinding stories of sly, burglary, and the extraordinary naughtiness that has procured these catlike characters a spot in the legends of networks all over the planet.

The Riddle of Feline Thieves

1.1 Characterizing Feline Thieves

The expression "feline thief" summons pictures of subtle lawbreakers exploring dull back streets and scaling structures with cat spryness. We investigate the starting points of the term and its association with the famous deeds of human robbers, making way for the catlike partners that would later become scandalous for their stealing ways.

1.2 Scandalous Felines in Mainstream society

Feline criminals have caught the creative mind of narrators and specialists, becoming notable figures in writing, film, and legends. From the secretive feline robber in an exemplary noir film to the perky vivified feline characters, we look at how these notorious cats have saturated mainstream society.

The Legends of Famous Felines

2.1 Tibbles the Feline: A Victorian Gem Cheat

In the Victorian period, one feline acquired reputation as a gem cheat in the clamoring city of London. Tibbles, a smooth and shrewd cat, turned into the focal point of consideration for his trying heists and capacity to steal valuable pearls. We unwind the legends encompassing Tibbles and his baffling ventures.

2.2 Felonious Cats of the Thundering Twenties

The Thundering Twenties saw a flood in wrongdoing, and a few cat partners joined the positions of famous characters. We dive into the stories of felines engaged with insignificant robberies, speakeasy antics, and the naughty soul of the Jazz Age.

Prominent Feline Criminals All over the Planet

3.1 The Feline Criminal of Amsterdam

In the channels of Amsterdam, a catlike cheat procured a standing for taking from neighborhood organizations and families. We investigate the narratives of this famous feline robber, inspecting the effect on the local area and the innovative endeavors to control its stealing tries.

3.2 The Parisian Pilferer: Le Visit Volant

Paris, the city of lights, turned into the setting for an infamous feline criminal known as Le Talk Volant, or the Flying Feline. This secretive cat slinked the housetops of the city, drumming up some excitement among occupants and policing. We disentangle the stories of this airborne instigator.

The Specialty of Cat Pilferage

4.1 The Brain research of Feline Criminals

What spurs felines to participate in stealing ways of behaving? We investigate the brain science behind cat pilferage, taking into account factors like impulse, liveliness, and the impact of the climate. Understanding the attitude of feline criminals adds profundity to the accounts encompassing their famous deeds.

4.2 The Infamous Feline Groups

Now and again, felines unite to make notorious feline packs, taking part in composed burglary and causing anarchy in their networks. We look at cases where gatherings of cat reprobates became amazing for their cooperative shenanigans and the difficulties looked by those endeavoring to ruin their exercises.

Underhandedness in Films

5.1 Energized Jokes: Tom and Jerry

The universe of movement has rejuvenated notorious felines in extraordinary ways. We investigate the devilish competition among Tom and Jerry, two notorious characters who have engaged crowds with their amusing and some of the time trying adventures. The enlivened couple's shenanigans exhibit the persevering through allure of cat wickedness in movies.

5.2 Aristocats and O'Malley's Misfortunes

In Disney's "The Aristocats," the cajoling stray feline O'Malley adds a dash of tricky appeal to the story. We dig into O'Malley's misfortunes and the depiction of felines as both enchanting and crafty in vivified films, adding to the ageless appeal of cat wickedness in true to life narrating.

The Scholarly Tradition of Wicked Felines

6.1 Puss in Boots: The First Joke artist Feline

The exemplary story of "Puss in Boots" includes a shrewd and creative feline who utilizes his finesse to raise his lord's economic wellbeing. We investigate the abstract tradition of Puss in Boots and how this prankster feline model has persevered in narrating, making a permanent imprint on the universe of writing.

6.2 The Cheshire Feline's Puzzling Smile

Lewis Carroll's "Alice's Undertakings in Wonderland" presents the baffling Cheshire Feline, known for its naughty smile and capacity to show up and vanish freely. We disentangle the imagery and tradition of the Cheshire Feline, inspecting job as a scholarly figure encapsulates the eccentric idea of cat naughtiness.

Web Sensations: Images, Viral Recordings, and Famous Felines

7.1 Cranky Feline: The Sovereign of Web Image dom

Cranky Feline, with her never-endingly disappointed articulation, turned into a web sensation, enthralling millions with her images and viral recordings. We investigate the peculiarity of Cranky Feline, looking at how her crotchety persona transformed into a social symbol and an image of web humor.

7.2 Nyan Feline and the Mainstream society Remix

Nyan Feline, a pixelated cat with a Pop-Tart body leaving a rainbow trail, turned into a viral sensation, motivating remixes, images, and even product. We dig into the social effect of Nyan Feline and its job in the web's interest with idiosyncratic and entertaining cat characters.

The Morals of Cat Naughtiness

8.1 Adjusting Naughtiness and Obligation

As we praise the devilish idea of felines, questions emerge about the moral contemplations of cat naughtiness, particularly when it includes burglary or possible mischief. We look at the moral components of appreciating and empowering perky cat ways of behaving while at the same time guaranteeing dependable pet possession.

8.2 Tending to the Outcomes of Feline Robbery

While stories of feline criminals might engage, there are genuine ramifications for both the felines and their human partners. We investigate examples where cat underhandedness prompted serious repercussions, provoking conversations about mindful pet consideration, local area mindfulness, and tracking down others conscious answers for forestall future episodes.

Feline Criminals Turned VIPs

9.1 The Reclamation of Notorious Felines

At times, feline thieves have become big names by their own doing, changing from infamous considers along with cherished characters. We inspect stories where scandalous felines tracked down recovery through restoration, reception, and endeavors to channel their devilish energy into positive commitment.

9.2 The Tradition of Infamous Felines

The tradition of infamous felines reaches out past their singular stories, impacting how society sees and values cat naughtiness. We ponder the enduring effect of notorious felines on mainstream society, writing, and the aggregate creative mind, recognizing their position in the more extensive account of human-cat connections.

Cat Naughtiness in Fables and Strange notions

10.1 The Joke artist Paradigm in Legends

Old stories from different societies frequently includes comedian figures, and felines, with their devilish nature, typify this prime example. We investigate how cat underhandedness is woven into folktales and customs, adding to the diverse imagery of felines in worldwide folklore.

10.2 Notions Encompassing Devilish Felines

From the beginning of time, strange notions have encircled felines, especially those known for their wicked ways of behaving. From convictions about felines taking one's breath to stories of enchanted changes, we dive into the notions related with famous felines and their effect on social insights.

The Feline's Whimper: Cat Underhandedness in Day to day existence

11.1 Embracing the Fun loving Soul of Felines

While stories of infamous felines might be loaded up with underhandedness and interest, there's a lighter side to cat shenanigans. We investigate the regular snapshots of fun loving nature that felines bring into the existences of their human partners, cultivating satisfaction, chuckling, and a more profound association between species.

11.2 Exploring the Complicated Relationship with Devilish Felines

Possessing a wicked feline accompanies its difficulties and prizes. We offer bits of knowledge into exploring the perplexing connection with cat miscreants, tending to the requirement for figuring out, persistence, and mindful consideration to guarantee an amicable concurrence among felines and their human partners.